EMPOWERING ARTIFICIAL INTELLIGENCE THROUGH MACHINE LEARNING

New Advances and Applications

EMPOWERING ARTIFICIAL INTELLIGENCE THROUGH MACHINE LEARNING

New Advances and Applications

Edited by

Nedunchezhian Raju, PhD

M. Rajalakshmi, PhD

Dinesh Goyal, PhD

S. Balamurugan, PhD

Ahmed A. Elngar, PhD

Bright Keswani, PhD

First edition published 2022

Apple Academic Press Inc.
1265 Goldenrod Circle, NE,
Palm Bay, FL 32905 USA

4164 Lakeshore Road, Burlington,
ON, L7L 1A4 Canada

CRC Press
6000 Broken Sound Parkway NW,
Suite 300, Boca Raton, FL 33487-2742 USA

2 Park Square, Milton Park,
Abingdon, Oxon, OX14 4RN UK

Library and Archives Canada Cataloguing in Publication

Title: Empowering artificial intelligence through machine learning : new advances and applications / edited by Nedunchezhian Raju, PhD [and five others].
Names: Raju, Nedunchezhian, editor.
Description: First edition. | Includes bibliographical references and index.
Identifiers: Canadiana (print) 20210344326 | Canadiana (ebook) 20210344350 | ISBN 9781771889308 (hardcover) | ISBN 9781774638125 (softcover) | ISBN 9781003055129 (ebook)
Subjects: LCSH: Machine learning. | LCSH: Artificial intelligence.
Classification: LCC Q325.5 .E47 2022 | DDC 006.3/1—dc23

Library of Congress Cataloging-in-Publication Data

Names: Raju, Nedunchezhian, editor. | Rajalakshmi, M., editor. | Goyal, Dinesh, 1976- editor. | Balamurugan, S., Prof. editor. | Elngar, Ahmed A., editor. | Keswani, Bright, editor.
Title: Empowering artificial intelligence through machine learning : new advances and applications / edited by Nedunchezhian Raju, PhD, M. Rajalakshmi, PhD, Dinesh Goyal, PhD, S. Balamurugan, PhD, Ahmed Elngar, PhD, Bright Keswani, PhD.
Description: First edition. | Palm Bay, FL, USA : Apple Academic Press Inc.; Boca Raton, FL, USA : CRC Press, 2022. | Includes bibliographical references and index. | Summary: "This new volume, Empowering Artificial Intelligence Through Machine Learning: New Advances and Applications, discusses various new applications of machine learning, a subset of the field of artificial intelligence. Artificial intelligence is considered to be the next big-game changer in research and technology. The volume looks at how computing has enabled machines to learn, making machines and tools become smarter in many sectors, including science and engineering, healthcare, finance, education, gaming, security, and even agriculture, plus many more areas. Topics include techniques and methods in artificial intelligence for making machines intelligent, machine learning in healthcare, using machine learning for credit card fraud detection, using artificial intelligence in education using gaming and automatization with courses and outcomes mapping, and much more. The book will be valuable for professionals, faculty, and students in electronics and communication engineering, telecommunication engineering, network engineering, computer science and information technology"-- Provided by publisher.
Identifiers: LCCN 2021049656 (print) | LCCN 2021049657 (ebook) | ISBN 9781771889308 (hbk) | ISBN 9781774638125 (pbk) | ISBN 9781003055129 (ebk)
Subjects: LCSH: Machine learning. | Artificial intelligence--Industrial applications.
Classification: LCC Q325.5 .E46 2022 (print) | LCC Q325.5 (ebook) | DDC 006.3/1--dc23/eng/20211201
LC record available at https://lccn.loc.gov/2021049656
LC ebook record available at https://lccn.loc.gov/2021049657

ISBN: 978-1-77188-930-8 (hbk)
ISBN: 978-1-77463-812-5 (pbk)
ISBN: 978-1-00305-512-9 (ebk)

About the Editors

Nedunchezhian Raju, PhD, is Professor in the Department of Computer Science and Engineering, Coimbatore Institute of Technology, Coimbatore, India. He has served in various capacities as Head of the Department, Vice Principal, Head of the Institution, and Director Research. His research interests include data mining, big data analytics, machine learning, and software Engineering. He has more than 25 years of experience in teaching and research. He has guided 18 PhD scholars, and six more PhD scholars are now doing their research under his supervision. To his credit, he has published more than 100 papers in refereed journals and international conferences. He has published a few books and book chapters also. He did his PhD, ME, and BE degrees all in Computer Science and Engineering.

M. Rajalakshmi, PhD, is Associate Professor at Coimbatore Institute of Technology, Coimbatore, India, in the Department of Information Technology. Her research areas include data structures, data mining, distributed systems, big data analytics, and machine learning. She has 25 years of teaching and research experience. She has guided four PhD scholars and is currently guiding 10 PhD scholars. She has published many papers in refereed international journals and international conferences and is a reviewer for few journals also. She did her PhD in Computer Science and Engineering at Anna University, Chennai, and she obtained ME Computer Science and Engineering degree from P.S.G. College of Technology, Coimbatore, India, with distinction. She did her bachelor's degree in Computer Science and Engineering at Tamil Nadu College of Engineering, Coimbatore.

Dinesh Goyal, PhD, has recently joined Poornima Group as Principal and Director at the Poornima Institute of Engineering & Technology, India. He is a renowned academician and a very good researcher. Dr. Goyal was previously Professor of Research & Academic Experience at Suresh Gyan Vihar University, Jaipur, India, for last 15 year, and has retired from the post of Director, Centre for Cloud Infrastructure and Security. He has expertise in information security, image processing, and cloud computing and has written more than 60 international and national papers. He has guided more than 20 postgraduate research scholars and as well as several research scholars. Dr. Goyal completed his bachelor's degree in Computer Science & Engineering from MBM College of Engineering, Jodhpur, India. He joined Rajasthan Technical University to complete his MTech with research work on "Information Security Activities and Societies" and was awarded the degree of PhD with specialization in Computer Science & Engineering from Suresh Gyan Vihar University, Jaipur, India.

S. Balamurugan, PhD, is Director of Research and Development at Intelligent Research Consultancy Services (iRCS), Coimbatore, Tamil Nadu, India. He has published 45 books, over 200 papers in international journals/ conferences, and 35 patents. He is the recipient of a Rashtriya Vidhya Gourav Gold Medal Award and a Best Educationalist Award by the Indian Solidarity Council, 2018. He is the recipient of two Lifetime Achievement Awards, one in 2018 and another in 2019. He is the recipient of Dr. A.P.J. Abdul Kalam Sadhbhavana Award by the International Business Council, India, 2018. He was awarded the Prestigious Mahatma Gandhi Leadership Award at the British Parliament, London, by NRI Welfare Society of India, 2019. The book he authored, Machine Learning and Deep Learning Algorithms using MATLAB and PYTHON, won a "Best MATLAB Book for Beginners" award from Book Authority, 2019. He won the CSI Young IT

Professional Award, 2017. His professional activities include roles as book series editor, editor-in-chief, associate editor, and editorial board member for more than 500+ international journals and conferences of high repute and impact. He has been invited as chief guest/resource person/keynote plenary speaker at many reputed universities and colleges at national and international levels. His research interests include artificial intelligence, soft computing, augmented reality, internet of things, big data analytics, cloud computing, and wearable computing. He is a life member of IEEE, ACM, ISTE, and CSI. He received his BTech, MTech, and PhD degrees in the field of Information Technology.

Ahmed A. Elngar, PhD, is the Founder and Head of the Scientific Innovation Research Group (SIRG) and Assistant Professor of Computer Science at the Faculty of Computers and Artificial Intelligence, Beni-Suef University, Egypt. Dr. Elngar is a Director of the Technological and Informatics Studies Center (TISC), Faculty of Computers and Information, Beni-Suef University. He has more than 25 scientific research papers published in prestigious international journals and several books covering such diverse topics as data mining, intelligent systems, social networks, and smart environments. Dr. Elngar is a collaborative researcher. He is a member of the Egyptian Mathematical Society (EMS) and International Rough Set Society (IRSS). His other research areas include Internet of Things (IoT), network security, intrusion detection, machine learning, data mining, artificial intelligence. big data, authentication, cryptology, healthcare systems, and automation systems. He is an editor and reviewer of many international journals around the world. Dr. Elngar won several awards including the "Young Researcher in Computer Science Engineering" from Global Outreach Education Summit and Awards 2019, as well as a Best Young Researcher Award (Male) (Below 40 years), Global Education and Corporate Leadership Award (GECL-2018), etc. Also, he holds intellectual property rights on "El Dahshan Authentication Protocol," Information Technology Industry Development Agency (ITIDA), Technical Report, 2016. Dr. Elngar's many activities in community and the environment service include organizing 12 workshops hosted by universities in almost all governorates of Egypt. He also organized a workshop on smartphones' techniques and their role in the development of visually impaired skills in various walks of life. Dr. Elngar received his PhD from the Faculty of Computers and Artificial Intelligence. Computer Science Department Beni-Suef University, Egypt.

Bright Keswani, PhD, is Professor and Head of the Department of Computer Applications, and Principal (Academic Staff College) and Editor-In-Chief (SGVU-Journal of Engineering & Technology) at Suresh Gyan Vihar University, Jaipur, India. He has a long standing of teaching at graduate and postgraduate levels for more than 16 years at various institutions. Presently, he is devoting his time to research in the varied field of computer science and giving research guidance to several MTech. and PhD students. Dr. Keswani is also associated with several companies and providing consultancy on project planning and designing to teams of software professionals. He has a number of research publications to his credit which have appeared in leading journals, some of which have been presented at international/national conferences and included in conference proceedings, leading international/national level magazines, and in-house journals of corporate sector. He is also of several international technical societies of Singapore, Canada, Australia, London, Hong Kong, Belgium, and the United States. He has held several positions for the Computer Society of India, Jaipur chapter. He is also a reviewer for various international journals of computer application in engineering, technology, and science and is a member of the editorial board of the International Journal of Computer Applications in Engineering, Technology and Sciences (IJ-CA-ETS). As a prolific writer in the arena of computer sciences and information technology, he penned a number of books. He has also authored self-learning material in computer science for several universities. Dr. Keswani was honored with a Best Citizen of India Award 2013, Academic Excellence Award 2015, Shiksha Bhushan Award 2009, and Bharat Jyoti Award 2013 for outstanding achievements in the field of computer science.

Contents

Contributors

Aafreen Nawresh
Department of Computer Science, University of Madras, Chennai, India

Jeelani Ahmed
Department of CS and IT, Maulana Azad National Urdu University, Hyderabad, India

Muqeem Ahmed
Department of CS and IT, Maulana Azad National Urdu University, Hyderabad, India

Nitin Arora
Department of Informatics, School of Computer Science,
University of Petroleum and Energy Studies, Dehradun, Uttarakhand, India

Alanknanda Ashok
Department of Electrical Engineering, Women Institute of Technology, Dehradun, Uttarakhand, India

S. Balamurugan
QUANTS IS & CS, Coimbatore, India

Devesh K. Bandil
Suresh Gyan Vihar University, Jagatpura, Jaipur

E. Chandra
Department of Computer Science, Bharathiar University, Coimbatore 641046, India

M. Chinnadurai
E.G.S. Pillay Engineering College, Nagapattinam, Tamil Nadu, India

Dharmaiah Devarapalli
Department of Computer Science and Engineering, Shri Vishnu Engineering College for Woman,
Bhimavaram, Andhra Pradesh, India

Kamlesh Dutta
National Institute of Technology, Hamirpur, India

Ahmed A. Elngar
Faculty of Computers and Artificial Intelligence, Beni-Suef University, Beni-Suef City,
Salah salem str., 62511, Egypt

Vikas Garg
Director, Executive Programs Management, Amity University Uttar Pradesh India,
Greater Noida Campus, India

Aditya Kakde
Department of Systemics, School of Computer Science, University of Petroleum and Energy Studies,
Dehradun, Uttarakhand, India

Bright Keswani
Department of Computer Applications, Suresh Gyan Vihar University, Jaipur, Rajasthan, India

Poonam Keswani
Akashdeep PG College, Jaipur, Rajasthan, India

S. Manikandan
E.G.S. Pillay Engineering College, Nagapattinam, Tamil Nadu, India

Mamta Martolia
Department of Computer Science and Engineering, Uttarakhand Technical University, Dehradun, Uttarakhand, India

Richa Mathur
C-404, Aarshiwad, Anandam, Balaji Market, Shrinathpuram, Kota, Rajasthan

Ambarish G. Mohapatra
Electronics and Instrumentation Engineering, Silicon Institute of Technology, Bhubaneswar, Odisha, India

T. V. Prasad
Visvodaya Technical Academy, Kavali, India

S. Sasikala
Department of Computer Science, University of Madras, Chennai, India

Durgansh Sharma
Department of Cybernetics, School of Computer Science,
University of Petroleum and Energy Studies, Dehradun, Uttarakhand, India

Dhanesh Kumar Solanki
E 325 III-B Khetri Nagar, Jhunjhunu, Rajasthan

Phanigrahi Srikanth
Department of Computer Science and Engineering, GMR Institute of Technology, Rajam, Andhra Pradesh, India

Shalini Srivastav
Commerce and Finance, Amity University Uttar Pradesh India, Greater Noida Campus, India

Poonam Tanwar
Department of CSE, Manav Rachna International University, Faridabad, India.

Dhananjaya Verma
Department of Computer Science & Engineering, ABES Engineering College, Ghaziabad, India

E. Udayakumar
Department of ECE, KIT-Kalaignarkarunanidhi Institute of Technology, Coimbatore, India

P. Vetrivelan
Department of ECE, PSG Institute of Technology and Applied Research, Coimbatore, India

Prity Vijay
Suresh Gyan Vihar University, Jaipur, Rajasthan, India

Vikash Yadav
Department of Computer Science & Engineering, ABES Engineering College, Ghaziabad, India

Abbreviations

AI	artificial intelligence
AN	assertion networks
AN	associative net
ANFIS	adaptive neuro-fuzzy interference system
ANN	artificial neural networks
API	application programming interfere
ASR	automatic speech recognition
BDA	big data
BDA	big data analytics
BDNF	brain-derived neurotrophic factor
BN	Bayesian network
BFS	breadth-first search
CAGR	compound annual growth rate
CD	conceptual dependency
CM	content management
CNN	convolutional neural network
CP	cerebral palsy
CT	computed tomography
DM	data mining
DBN	deep belief network
DFS	depth-first search
DN	definitional networks
DRNN	deep recurrent neural network
EF	ejection fraction
EHRs	electronic health records
ELU	exponential linear unit
EMR	electronic medical record
EN	executable networks
ESPMS	expert system for process model selection
FD	false detection
FIELD	Fanuc Intelligent Edge Link and Drive System
FIS	fuzzy inference system
FKRS	formal knowledge representation system

FN	false negative
FP	false positive
FR	false rate
GANs	generative adversarial networks
GDP	gross domestic product
HN	hybrid networks
HKR	hybrid knowledge representation
HPRT	hybridization preprocessing and resampling technique
II	information integration
IoT	Internet of things
IR	information retrieval
IN	implicational networks
IVS	inter ventricular septum
KA	knowledge acquisition
KB	knowledge base
KR	knowledge representation
LA	left atrium
LCD	liquid crystal display
LED	light-emitting diode
LN	learning networks
LSE	least square estimate
LV	left ventricular size
LVPW	left ventricular posterior wall
MFs	membership functions
ML	machine learning
MRI	magnetic resonance imaging
NHANES	National Health and Nutrition Examination Survey
NLP	natural language processing
NLP	processing of natural language
NN	neural network
NSG	next-generation sequencing
OBLOG	object-oriented logic
OS	oversampling
PSN	partitioned semantic net
PL	predicate logic
PHI	personal health information
QAS	question answering system
RAISE	Responsible AI for Social Empowerment

RBS	rule based system
RDD	resilient distributed dataset
RDF	resource description framework
RELU	rectified linear units
RNN	recurrent neural network
ROP	retinopathy of prematurity
SDGs	sustainable development goals
SNs	semantic networks
STEM	Science, Technology, Engineering, and Math
SVMs	support vector machines
T2DM	type-2 diabetes mellitus
TN	true negative
TFP	total productivity factor
TP	true positive
US	undersampling
VR	virtual reality
WHO	World Health Organization
ZCR	zero crossing rate

Preface

Artificial intelligence (AI) is considered to be the next big game changer in technology. We are living in the time when every person in the world has access to technology. Technologies in the past few years have grown by leaps and bounds. Computing has enabled machines to learn, making machine and tools became smarter. They can now optimize and act as per the practices of the user and save a lot of time and money of the user. This is the sign of many upcoming advancements in application of AI in various sectors of science and engineering. Empowering machines to think and take decisions by themselves will make the machine able to enrich and enhance in knowledge and capacity. AI has seen its development since past half century, and today AI is capable to work faster, smarter, and more accurately than humans.

AI started its journey in 1956, when John McCarthy coined the term. A lot of research had been done in the past decades on AI and expert systems, its merits and demerits, and possible application areas. According to a Forrester research survey, 38% of industries adopted AI in 2017, and it is expected to reach 62% by the end of the calendar year 2018. It is also predicted in research that there is a marathon increase of 300% in investment in AI in 2017 as against 2016. According to research carried out by International Data Corporation (IDC), it is predicted that the growth of the AI market will rise from $8 billion in 2016 to $47 billion by 2020. Among the huge number of application areas of AI, Forresters analysis predicted 10 key technologies where there is a huge scope for application of AI.

The key technologies include:

- natural language generation speech recognition
- virtual agents
- machine learning platforms
- AI-optimized hardware
- decision management
- deep learning platforms
- biometrics
- robotic process automation

- text analytics
- natural language processing

According to a research survey conducted by BIN sights, AI is going to have its huge impact in the following sectors in the future: healthcare, advertising, sales and marketing, business intelligence, security, finance, IoT/wearables, education, customer relationship management, personal assistants e-commerce, and robotics.

Thus, the well laid down 12 chapters of this book discuss and analyze the existing and new technological advancements in the domain of machine intelligence and artificial learning.

The summary of these 12 chapters is as follows:

- Chapter 1: It takes a detailed survey on various knowledge representation techniques in AI for making machines intelligent, wherein all the techniques of knowledge representation and tools for creating these knowledge representation are being compared and analyzed.
- Chapter 2: This work focuses on ontology architecture to process different semantic information to increase the sentiment accuracy by implementing different advance machine learning techniques and algorithms.
- Chapter 3: This chapter discusses the emergence and applications of AI in healthcare by outlining different machine learning trends in the healthcare sector.
- Chapter 4: In this chapter, the author discusses two latest IT technologies that are big data analytics and Internet of Things and their applications in healthcare.
- Chapter 5: This research work addresses one of the best challenges to predict patients and behavior of the patients. A new framework GROCD addressed novel fuzzy rules with efficient clustering and classification of patients has been designed and proposed. This framework proposed control ambiguity of a specially reducing matrix. Also, this proposed framework gives accurate results followed as designed novel fuzzy rules with efficient clusters with proposed similarity measure, classification of KNN of classified patients using proposed similarity measure sigmoid with Euclidean space predicted patients.
- Chapter 6: In this work, the author hypothesizes that the traditional neural network algorithm does not fit for an imbalanced dataset. In

the second experiment, HPRT is applied on the dataset for reducing dataset complexity where the performances of neural network algorithm also boost up, bringing forth precise and accurate result.

- Chapter 7: This work proposes the use of AI in education using gaming and automatization with courses and outcomes mapping.
- Chapter 8: This chapter examines the economic and business impact of artificial intelligence for comprehensive growth across the globe.
- Chapter 9: This chapter proposes an architecture that will help to monitor plants by taking the images in real time and then classify it. The model implemented for this classification is based on a convolution neural network, and a novel approach has been used to get better performance for classification tasks.
- Chapter 10: This chapter discusses application of AI in image processing.
- Chapter 11: In this chapter, a new control system of a voice-controlled wheelchair using an AI algorithm is proposed. The machines are given instruction through the voice signals. It has become inevitable for communication in this current state of technology.
- Chapter 12: In this contribution, the authors propose an expectation framework for coronary illness, utilizing Adaptive Neuro Fuzzy Interference System. This framework acknowledges different clinical highlights as contribution as name, age, sex, blood pressure (systolic), aorta (Ao), blood pressure (diastolic), left ventricular size (LV), inter ventricular septum (IVS), ejection fraction (EF), tricuspid valve, left ventricular posterior wall (LVPW), left atrium (LA), and pulmonary velocity as input.

This edited book provides details of the various applications of AI in diverse disciplines and sectors. This book will serve as a primary source of reference for different applications of AI and machine learning.

CHAPTER 1

A Tour of Various Knowledge Representation Techniques in Artificial Intelligence for Making Machines Intelligent

POONAM TANWAR[1], T. V. PRASAD[2], and KAMLESH DUTTA[3]

[1]*Department of CSE, Manav Rachna International University, Faridabad, India*

[2]*Visvodaya Technical Academy, Kavali, India*

[3]*National Institute of Technology, Hamirpur, India*

[*]*Corresponding author. E-mail: Poonamtanwar.fet@mriu.edu.in, poonam.tanwar2012@gmail.com*

ABSTRACT

Two major things are required in order to make any computer or mechanical device or machine exhibit intelligence as a man's brain and they are to develop exhibit intelligence as human, that are knowledge representation (KR) and inference mechanism. Development of an artificial intelligence (AI) system has been an important job since at certain times incomplete information is available which can be vague and dicey. Hence, the answer to these concerns is to create effective database and an impactive inference mechanism. KR is an attractive zone spread over varied areas of cognitive science as well as computer science. It is difficult to recognize the need of an amalgamation of various techniques and conclusion mechanism to acquire the accurate results to the said problem domain. The basic model of the KR system has been explained in this chapter. This chapter also elaborately discussed the KR technology and various KR approaches with respect to different application. As we know, AI applications are very vast and can be applied in various domains like vision, problem solving, decision-making, robotics, question answering system (QAS) and natural language processing (NLP). The deep survey of core area of AI and various KR techniques is the objective of this chapter.

1.1 INTRODUCTION

Artificial intelligence (AI) (Russell and Norvig, 2009) incorporates intelligence of a human machine. Russell and Norvig, 2009 stated that AI is the branch of science which makes the machine exhibit intelligence as man's mind for a specific segment. Alan Turing in 1950s presented his work on computing machinery and intelligence (Sowa, 1992) and stated that Turing test can be used to test the machine intelligence. Hayes and McCarthy (1969) state that a machine is intellectual probably for reasons giving solution to a kind of situation needing human brain. Some more explanations/descriptions for AI were also expressed as "AI is the part of computer science concerned/related with designing computer systems that exhibit the characteristics we associate with intelligence in human behavior." Charniak and McDennott (1985) state that "AI is the study of mental faculties through the use of computational models," whereas Yousheng Tian et al. (2011) state that it is the study of cognitive science. Mylopoulos (1983) discussed a brief description on terminology and worked for open domains related to KR.

The survey is classified into KR, HKR, and other KR techniques

1.2 KNOWLEDGE REPRESENTATION

Problems in AI require (Tanwar et al., 2014) extensive amount of know how encompassing the environment of the world. Articles, object's attributes, their relations, classes, events, situations, causes states, and time and (Tanwar et al., 2014) impact are the parameters to be presented by AI. Existing KR techniques can be used to represent these things Brewster et al. (2004). The KR system performs the three major tasks: acquisitions, reasoning, searching (Fig. 1.1).

1.2.1 ACQUISITION

Rosenbloom and Rosenbloom (1988) states that the acquiring of knowledge can be seen as a method of acquiring shaped base of knowledge and can be a process of incorporating/integrating novel ideas into the knowledge base. In addition to KR, Inference and Retrieval Knowledge Acquisition is also an important part of developing an AI system. Knowledge acquisition (KA) not only means to add new knowledge to the KB, but it also requires collaborating together with new knowledge to one already in existence.

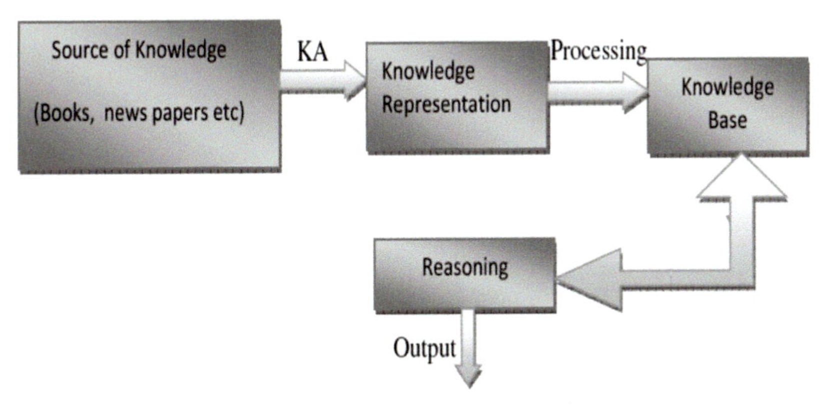

FIGURE 1.1 KR model.

It can be done by providing the relationship between the new knowledge element that the system requires and the already existing elements that is the knowledge base (KB). These relationships play an important role in accessing the knowledge in which KB is stored. Acquisition can be performed by two different ways.

a) The element can be linked directly with the existing one or
b) They can be connected through clusters.

KA is classified into the following categories by Bill and Gil (1996) and the comparison between them is explained in Table 1.1.

- Symbol-Level Approaches
- Role Limiting Approaches
- The Composable Role-Limiting Methods

1.2.2 REASONING

Reasoning is the process of generating new knowledge from the knowledge that a KB stores. A separate mechanism is used for inferencing the

TABLE 1.1 Various Knowledge Acquisition Methods/Tools for Text Data.

Category/tool	Symbol-level approaches	Role-limiting approaches	Composable role-limiting methods	Reasoning/control strategy	Applications
TEIREISIAS	✓			Rule chaining	Diagnostic system
MORE		✓		Rule based	Diagnostic system
SALT		✓		Methods composed of many library/manually by KA tool's designers	Natural language processing, problem solving
ROGET		✓		Rule based	Expert system
DOT	✓		✓	Form based	A KBS system that trouble shoots the DNA sequencing machine
PROTÉGÉ-II			✓	KBS builder manually select the required methods from the library	Problem solving
KRITON			✓	Hybrid of role and frame	Expert system

knowledge from existing KB. Unification and resolution are the methods of reasoning used in the first order predicate logic (PL). In case of semantic net, inheritance is the mechanism for the reasoning, purpose, and path-based reasoning can also be used. There are two methods in which rule-based system (RBS) applies the rule, see Tables 1.2 and 1.3. A strong control strategy is required for the flooding of the knowledge acquitted by initial state/staring point to the terminating point/goal state without ambiguity.

TABLE 1.2 Application of RBS.

Chaining	Method	
Forward chaining	The situation of the system is always paralleled along the ancient rule which is on the left side of the rule.	Action to be applied is the right side/conclusion of corresponding rule.
Backward chaining	The situation of the system is always compared with the right side/what could be the conclusion of the rule.	Action to be applied to ancient/ left side of that rule.

TABLE 1.3 Characteristics of Forward and Backward Chaining (Tanwar et al., 2014).

Forward chaining	Backward chaining
Will be applied as present to future	Will be applied as present to past
Planning, controlling and monitoring	Diagnosis system
Data driven and bottom-up reasoning	Goal driven
To follow from the facts	Support the hypotheses
Work forward to find what solutions	Work backward to find facts
Antecedents determine search	Resultant determine search
Breadth-first search facilitated	Depth-first search facilitated
Explanation not facilitated	Explanation facilitated

1.2.3 SEARCHING

There are various search techniques/algorithms available in AI mentioned by Russell and Norvig (2009) and Tanwar et al. (2010). These techniques are divided into two categories: blind search technique and heuristic search techniques, Rich and Knight (1991). In blind/uninformed search, no additional information is provided to search the goal to follow the path based on meaning embedded in the name of the search to reach the goal by generating of the every possible node/state. The depth-first search (DFS) and breadth-first search (BFS) lie in this category, whereas in Heuristic Search, additional information is provided, that is, a heuristic function that provides the optimal path by their possibility of success. Steepest hill climbing, best first search, and/or graphs, alpha, and beta pruning are examples of heuristic search.

1.3 KNOWLEDGE REPRESENTATION TECHNIQUES

KR methods are broadly divided in three categories: declarative KR techniques, procedural KR techniques, and hybrid KR techniques (Morgenstern, 1999) and (Mylopoulos, 1983). A declarative KR is defined as every piece/chunk of knowledge that allows the reasoning system/inference mechanism to use the rules of inference and to come out with a new piece of information. Whereas the procedural representation represents the knowledge as procedure, allows the inference system to manipulate the defined procedures to arrive at the result, Rich and Knight (1991) and Tanwar et al. (2011).

The HKR is a combination of any of the following:

- More than one declarative method can be used.
- More than one procedural method can be used.
- One or more procedural method can be combined with one or more declarative method.

1.3.1 DECLARATIVE AND PROCEDURAL KR TECHNIQUES

Declarative KR techniques have been applied for the knowledge which tells about something like, "Yash is studying in class 4." Declarative knowledge can be conscious, verbalized, metalinguistic, or knowledge about a linguistic form, whereas procedural KR techniques are used to represent how things can be done or how things tend to do like formatting of the computer.

Now days, many KR techniques are available for displaying knowledge like:

- Tree and graph effectively represent hierarchical knowledge
- Semantic networks (SNs) help to represent the knowledge using nodes and links which mainly store/describe the propositions (Stallings, 1994).
- Common sense knowledge can be represented by Schemas.

- Frames and scripts are the commonly used KR techniques presented by Rich and Knight (1991) and Russell and Norvig (2009). Frames describe the objects which can be represented by slots and their relationship can be represented by links. Each slot in the frame is organized hierarchically. Script writings are used to detail the stereo type/event rather than objects, Rich and Knight (1991). It consists of stereo-typically ordered causal or temporal chain of events. Declarative and procedural KR techniques are described in detail in this section.

1.3.1.1 LOGIC

Classified in two categories, that is, first-order logic (FOL)/ PL and propositional logic can be used for statements that are universally false or true, in contrast to the life-like situations, has many statements/concepts which can't be applied like this, for example, "all human love each other" (Rich and Knight, 1991). PL works for such kind of statements. FOL/ PL extends propositional logic in two parts, one side it provides an inner structure for sentences was provided. They can be seen as an expression of connect between things/objects or individuals Rich and Knight (1991) (Tanwar et al., 2012). Second, it establishes an expression of the knowledge along with an equation which generalizes (Russell and Norvig, 2009). It makes possible to answer that a kind of property takes on all objects, of some objects, or of no objects. Two additional notations are also defined in PL, that is, universal quantifier and existential quantifier. They are used to limit the scope of the variable (Rich and Knight, 1991).

1.3.1.2 SEMANTIC NET

Semantic Net (SN) is broadly used in KR methodology and is used to denote the link between things/concepts/objects/instants or categories/class of instances. It is an indicative graphical direct in which objects/class is represented by nodes/vertices and boundaries or linkage (unidirectional) denote the semantic relations among the class and objects (Tanwar et al., 2010). SN is used to represent the inheritable facts and inheritance can be used as an inference mechanism in SN. Inheritance is the property by which the objects/things of base class inherit the values and attribute from base class as shown in Figure 1.2. To give hold to the inheritance things be so organized that they fall into the classes which in turn must be aligned in a generalization hierarchical order (Shapiro, 1977) and (Tanwar et al., 2010).

Sometimes SN is also known as associative net (AN) as the nodes are in association with other nodes as there is a stimulation spreading from one concept/class node to other nodes (Tanwar et al., 2010 and Shapiro, 1977). These types of particular associations have been proven useful in a range of KR. The links that have been used in SN are IS-A and A-KIND-OF. IS-A means are part of or are in reference to a member belonging to some class whereas A-KIND-OF is used to represent the connection between one class to various class as shown in Figure 1.2.

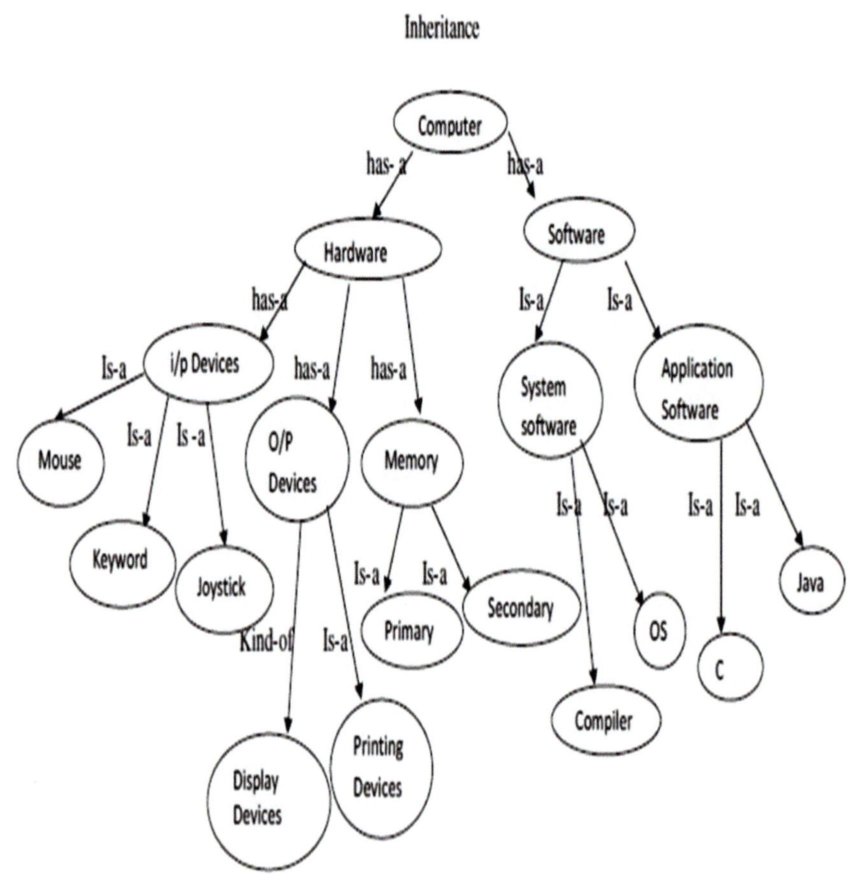

FIGURE 1.2 Property of inheritance.

TABLE 1.4 Parameters of Predicate Logic.

Notation	Description	Example(s)	Remarks
Terms	The First-order logic are used to represent objects or individuals. Terms can be a constant (designate specific object)	A, B, Smith, Blue, etc. variable (designate unspecified object): x, y, z, etc. and Functions (designate a specific object related in a certain way to another object or objects): Father Of, Color Of	
Predicates	A relation that binds two atoms have a value of true or false. A predicate can take arguments, which are terms. A predicate with one argument expresses a property of an object. A predicate with two or more arguments expresses a relation between objects	• Student (Bob) • Likes (Bob, Mary)	A Predicate without arguments is just simple proposition logic
Universal Quantifier	Used to identify the scope of the variable in logical expression	• $\forall$ x P (x) means "for all x, P of x is true" • $\forall$ x Happy (x) If the universe of discourse is people, then this means that everyone is happy	
Existential Quantifier	If the statement is $\exists$x P(x) means "there exists at least one x for which P of x is true"	• $\exists$x Happy (x), If the universe of discourse is people, then this means are is at least one happy person • $\forall$x $\exists$y Knows (x, y), ^ Knows (y, x). $\forall$x $\exists$y Knows (x, y). => $\neg$ Likes (y, x)	

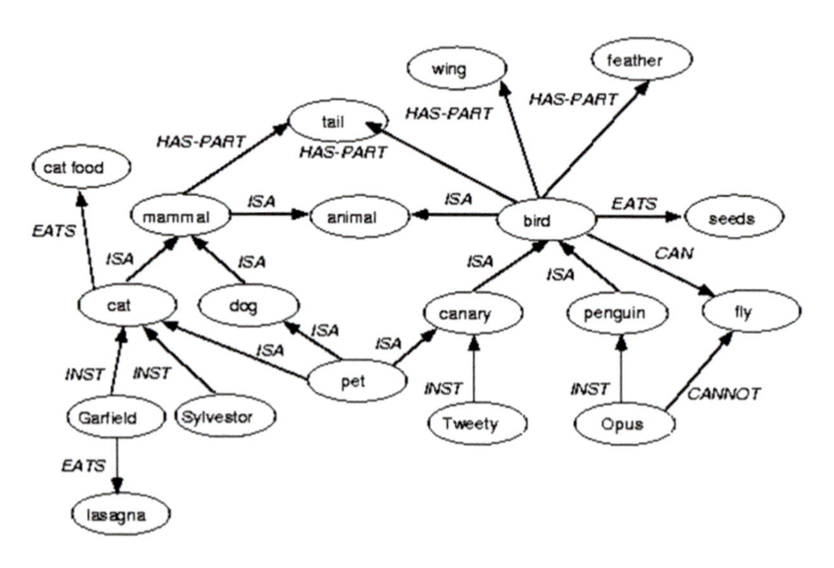

FIGURE 1.3 Representation of IS-A, HAS, INSTANCE (Shetty, 2006 and Tanwar et al., 2010).

SN is a declarative graphical representation which represents either knowledge or support-automated systems for reasoning about knowledge. The six most common kinds of SNs as described by Shapiro (1977) and used by Tanwar et al. (2010) are given below:

1. Definitional networks (DN)
2. Assertion networks (AN)
3. Implicational networks (IN)
4. Executable networks (EN)
5. Learning networks (LN)
6. Hybrid networks (HN)

Walker (1975) presented the partitioned semantic net (PSN) which came to limelight for the speech understanding system. Then later in 1977, Gary explained how one could expand the utility of semantic net using PSN (Frost, 1985).

PSN delimit the scopes of quantified variables mentioned by Gary (1977). As worked on the quantified statements, it is of great help to denote the very informative chunk consisting of any happening/event. Following is the example, "Poonam believes that the Earth is round" is shown in Figure 1.4. Node <POONAM> an agent of event node, whereas

<EARTH> and <ROUND> represent the objects of portioned network (Tanwar et al., 2011).

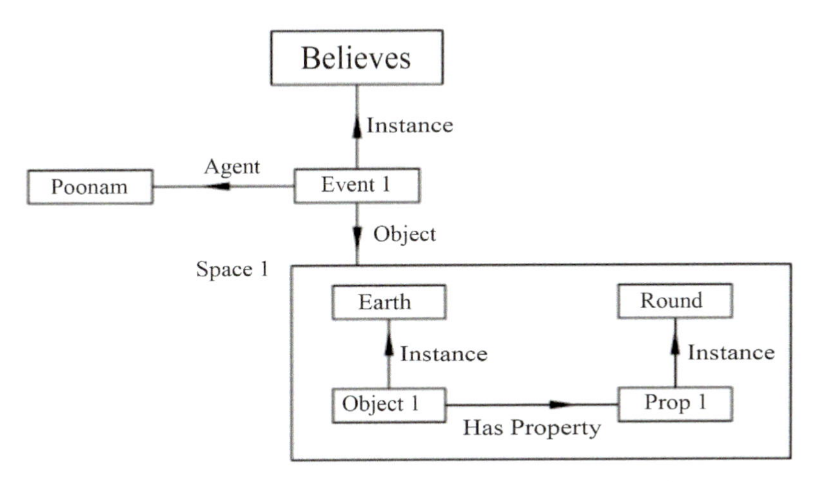

FIGURE 1.4 Partitioned semantic net (Gary, 1977; Tanwar et al., 2010).

PSN could also represent the quantifiers (Universal and Existential). For instance, "Every sister knots the Rakhee to her brother," can be easily represented in PL. Rakhee R and Sister S are indicative of objects while knot is expressed by a predicate, whereas in semantic net the event is representation of a thing/object of a class/complex objects, that is, knot represents a situation of more complex objects. Universal quantifier can be effectively represented by PSN. For instance, "Every sister knots the Rakhee to her brother" is depicted in Figure 1.5. They could also make use of complex quantifications which involve nested scope by use of nesting space (Tanwar et al., 2010).

1.3.1.3 FRAMES

The frame KR is an extension of SN as the complexity of KR increases as complex the knowledge. The frame KR will be helpful in such kind of situations. A frame is an attributive collection and their associative values like objects which give description of the real-world entity. Frame indicates, A class (set)/attributes/an instance (an element of a class)/object (Tanwar et al., 2011) and (Singhe et al., 2003).

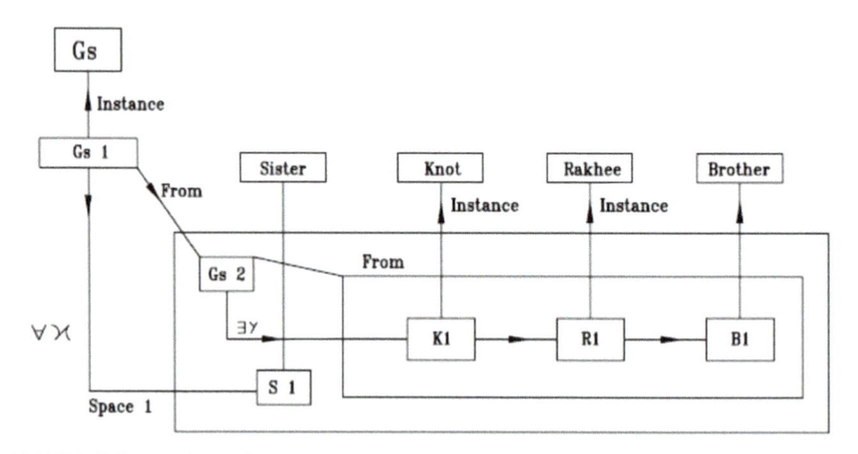

FIGURE 1.5 Partitioned Semantic Net for Quantifiers (Gary, 1977; Tanwar et al., 2010).

- Frame can be headed under basic three categories:
- Frame of name
- Attributes of frame (slots)
- Values of frame (fillers: list of values, range, string, etc.)

Two naming conventions for the frame are: first the true name with unique description of the frame and second, having various number of public names. For instance, Frame frame-20 will look as:

Title: ("Boy")
Marital Status: (frame-4)
Friend: (frame-20)

1.3.1.4 PRODUCTION SYSTEM

In 1979, Waterman and Hayes-Roth offered a procedural scheme via production rules that became most popular and widely used KR technique (Stallings, 1994). Production system consists of four components (Russell and Norvig, 2009):

- A set of rules/rule base each of which contains the left hand side that determined the applicability whereas right side that determines which operation has to be performed if the rule is applied.
- A KB that contains whatever information is required to solve given problem. It may be permanent or may be used for the current problems.

- A strategy of control will be specific to the structure to which the rules will be fired and also compare the rule with each knowledge stored in KB and a means of setting conflicting situation, rule applier/interpreter is used to decide which rule to execute on each selection execute cycle.
- Advantages:
  - Naturalness of expression
  - Modularity
  - Restricted syntax
  - Ability to represent uncertain knowledge
- Disadvantages:
  - Inefficient
  - Less expressive

1.3.1.5 CONCEPTUAL DEPENDENCY

Conceptual Dependency (CD) was proposed by Roger Schank in 1973 to express knowledge gained from natural language input. In CD, a series of diagrams represents sentences that depict the use of not concrete situations along with real physical situations. Primitive actions, different types of states, and different theories of inference can be represented by CD as shown in Figure 1.6. Two axioms it may have:

- Sentences that have identical meaning could have only one/single representation.
- All implicit information's has to be made explicit in the representation.

The CD defines precise usual primitives used to provide the meaning like objects, modifier of action shown in Table 1.5.

1.3.1.6 SCRIPTS

A chunk of series of events using the concept of structural object called scripts which was proposed by Roger Schank and his friends in 1977 (Stillings, 1994). It is not a passive information and has in it a category of happenings which are contextual, has participation and part of events which signify the slotted collection or sequential frames using inheritance and slots.

TABLE 1.5 Primitive Defined in CD.

Sl. No.	Primitive	Action
1	ATRANS	Transfer of an abstract relationship.example "take".
2	PTRANS	Transfer of the physical location of an object example Jump.
3	PROPEL	Application of a physical force to an object. Example pull.
4	MTRANS	Transfer of mental information. Example ask
5	MBUILD	Construct new information from old. Example decide
6	SPEAK	Utter a sound. Example say
7	ATTEND	Focus a sense on a stimulus. Example listen, watch
8	MOVE	Movement of a body part by owner Example punch, kick
9	GRASP	The Actor is grasping an object. Example clutch
10	INGEST	The Actor ingesting an object. Example, eat
11	EXPEL	The Actor is getting rid of an object from the body

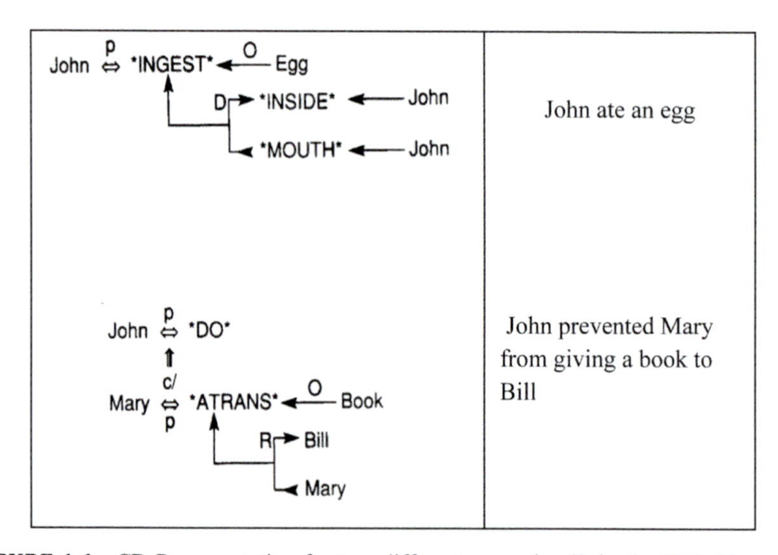

FIGURE 1.6 CD Representation for two different examples (Schank, 1973; Tanwar et al., 2011; 2013).

For example, "Rohan went to the restaurant and had some pastries." The meaning derived from the above text shows how he gets the pastries from the restaurant for eating. Though no such information is given in the text, script defines an episode with the known behavior and describes the sequence of events as shown in Figure 1.6. If any new condition is encountered, then the new information is accepted with the desired exception. Scripts could contain other information as well, which include the following mentioned by Davis et al. (1993) and used by Tanwar et al. (2011), broadly analyzing how the script behaves in a given circumstance.

One of the most common examples of the script is the restaurant shown in Figure 1.7.

Script: RESTAURANT			
Track:	Coffee Shop	Entry cond.:	S hungry
Props:	Tables		S has money
	Menu		
	F=Food	Results:	S has less money
	Check		O has more money
	Money		S is not hungry
			S is pleased (optional)
Roles	S=Customer		
	W=Waiter		
	C=Cook		
	M=Cashier		
	O=Owner		

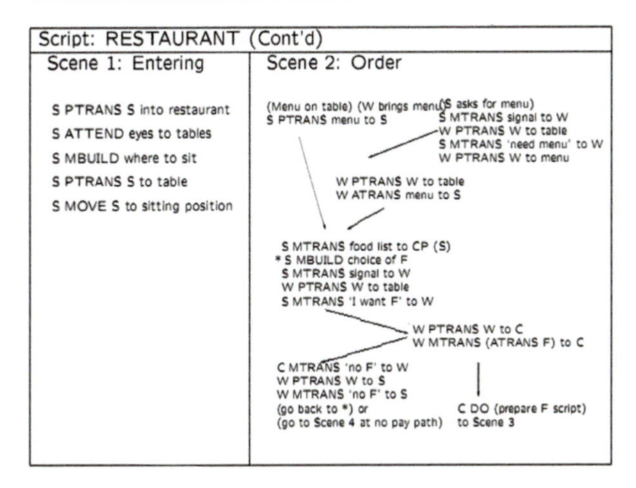

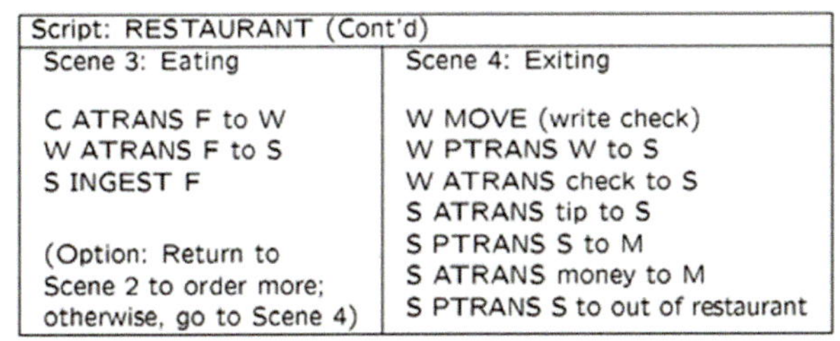

FIGURE 1.7 Script structure for restaurant (Davis et al., 1993; Tanwar et al., 2011; 2013).

As in the above case, PTRANS represents the transfer of physical location of the object ATRANS is utilized for the transfer of the relationship.

1.3.1.7 ANALOG

In 1994, Syed S. Ali presented a new KR technique for NLP. The syntax of the A Natural Logics (ANALOG) is a language designed for propositional SN representation formalism used for NLP. ANALOG involves the syntax and semantics of the logic of structured literals. The advantages of structure variables in representation are subsumption described by Syed Ali (1997). The deduction of subsumption from the graphical representation of preposition (with the help of nodes and edges) was based on the subsumption mechanism of KL ONE (1985) and Krypton (1985). The subsumption could be defined syntactically, in terms of the structure of the structured variables. For example, from a rule representing, "every boy loves a girl," all rules involving more restricted boys follow directly by the variable subsumption (e.g., "every boy that owns a red car loves a girl") and the matching process of variable to specific class is shown in Figure 1.8. This kind of subsumption can be defined without the traditional T-Box and A-Box distinction and has a great deal of utility. Each node in the system called as variable node and semantics and syntax of logic of representation was based on the system developed by Morgado in 1986 and Shapiro in 1991.

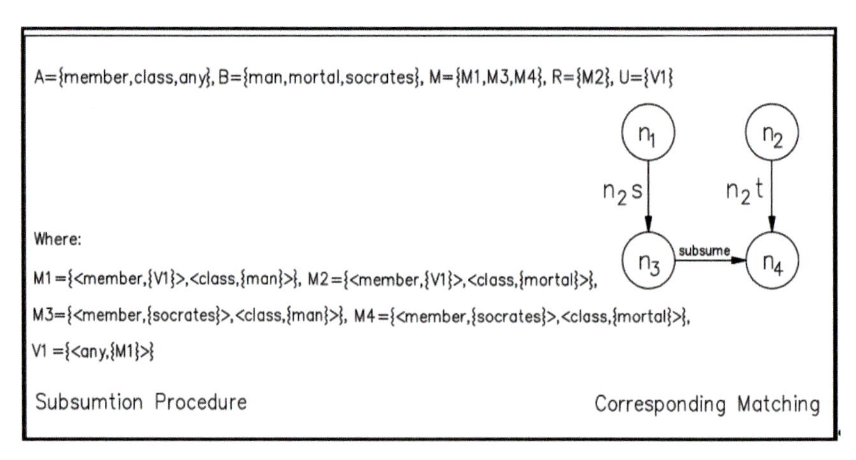

FIGURE 1.8 Subsumption and matching process used in Analog Ali, 1994; Tanwar et al., 2018).

1.3.1.8 ONTOLOGIES

Davis et al. (1993) proposed the ontology which is used to represent real-world entities and arrange them in a hierarchy shown in Figure 1.8. Ontology further categorizes the entities based on their similarities presented by Davis et al. (1993). KR by ontology can be easily accessible by the machines. It binds the people/entities in a community based on the conceptualization. Domain-specific knowledge can also be represented by ontology. Symbols used in this KR are mainly used to represent the perceptions/concepts and their relations (Costa, 2014). Ontology can be used in wide applications like content management (CM), information integration (II), architecture engineering and construction, and information retrieval (IR).

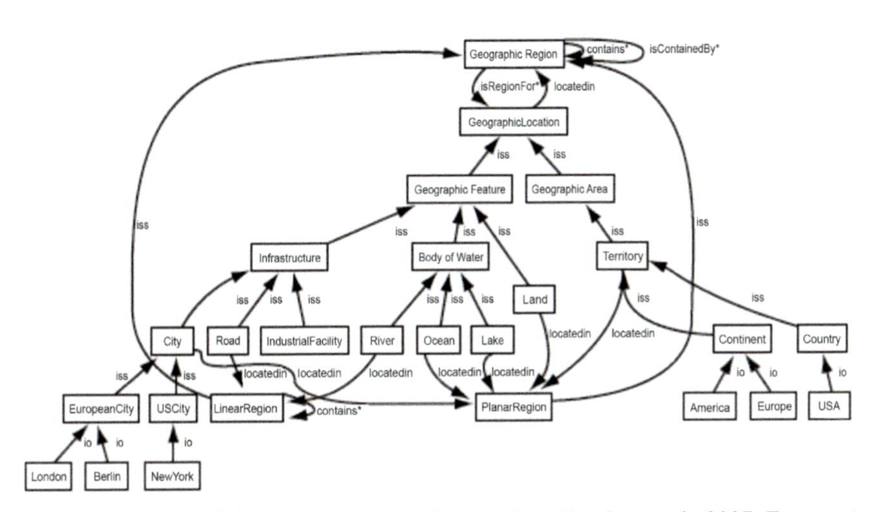

FIGURE 1.9 Knowledge representation using ontology (Stephan et al., 2007; Tanwar et al., 2018).

1.3.1.9 BAYESIAN NETWORK

Uncertain knowledge can be represented by Bayesian Network (BN) Nodes are used to represent the number of variables, $Z = Z_1, Z_2, Z_3 \ldots\ldots Z_n$ and they can be joined by directed arcs/links. The connected nodes represent the dependencies between variables, are used to represent the propositions, and can have ordered value. The descriptive reasoning in

BN is from symptoms to cause and predictive reasoning is reasoning about causes to the new belief mentioned by Zhang and Guo (2006).

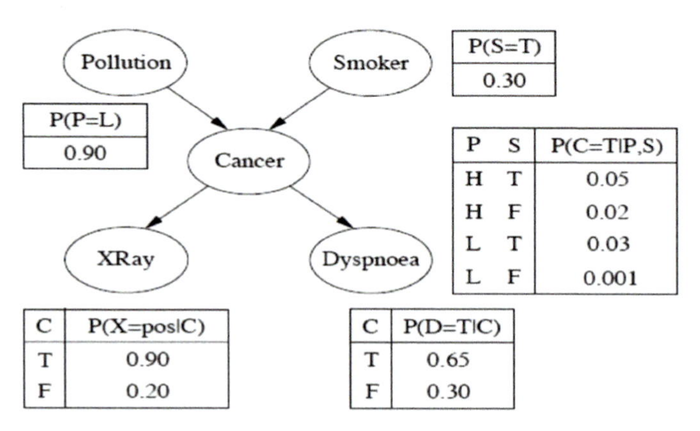

FIGURE 1.10 Bayesian network for domain lung cancer (Kevin and Nicholson, 2010 and Tanwar et al., 2018).

1.3.1.10 SEMANTIC WEB

Semantic web technology can be used to represent KR on web. Semantic web is an extension of the current one, with well-defined syntax and semantics or we can say that meaning of information which enabled computers and people to work in cooperation presented by Heflin (2001). Semantic web is used to record the data and help to represent the data in such a way so that it resembles with the real-world entity. Semantic web KR helps in data integration, management, conceptualization, and retrieval.

1.3.2 HYBRID KNOWLEDGE REPRESENTATION (HKR) TECHNIQUES

As per user need to overcome the disadvantages of simple KR techniques and strengthen the representation capability of single KR system research in KR come up with various HKR techniques/system like FRORL Krypton, LOOM, Mantra & Extended Semantic Web, etc. The comparison of various KR tools are given in Table 1.6.

In fact, KR is the most critical and fundamental issue in AI that attempts to understand intelligence, Eric (2007) and Jain and Mishra (2014). The main problem of AI system is how to represent knowledge and how to incorporate both types of knowledge in a single system (declarative and procedural). Due to these problems, KR became a separate research area in AI. For a few years, a group of systems has evolved that incorporate all these fundamental issues known as the HKR system, which essentially includes two or more KR techniques.

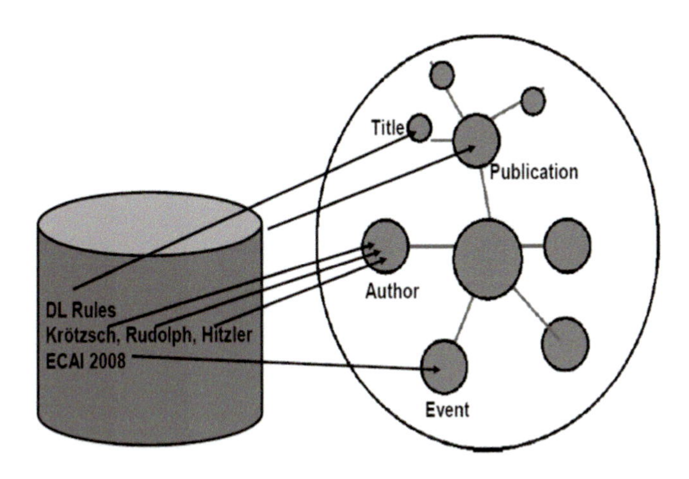

FIGURE 1.11 Knowledge to be represented using semantic web (Hitzler, 2011; Tanwar et al., 2018).

Implementation of HKR started from 1980 mentioned by James et al. (1983), with KL-ONE which was based on PL frame. It was a powerful HKR tool but only used for representing declarative knowledge. Krypton, a TELL-ASK module developed by Ronald et al. (1983) was based on KL-ONE again used for declarative KR. Different HKR tools were investigated for different application areas like KL-ONE, Krypton, object-oriented logic (OBLOG); too many tools were proposed and implemented. Quesgen and Voss (1987) presented the Babylon HKR tool which was implemented for procedural knowledge but the system required prior knowledge to apply to query. Calmet et al. (1991) proposed the modular assertional, SN, and terminological representation known as MANTRA. HKR Tool which was implemented in 1991 with the graphical KB used for representing the procedural knowledge was used to design a model for mathematical KR, appropriate for computer algebra systems. FRORL developed by Jeffrey (1992) for designing the prototype model for software, AAANTS developed by Chirminda et al. (2003) to design the intelligent environment and Reena et al. (2006) proposed the extended semantic net for document classification and indexation. HKR in 2010 was used to implement the mathematical model for algebraic problems by Nhon and Nguye (2010). In 2011, formal knowledge representation system (FKRS) was developed for autonomous concept formation based on conceptual algebra. Expert system for process model selection (ESPMS) HKR was used to build the decision-making system for software engineers mentioned by Ismail and Rehman (2011).

The reason behind the implementation of HKR system was its day-to-day growing usages/applications like robotics, expert systems, medical diagnostics, NLP, IR, machine learning, and other applications of AI, cognitive science, and in all the areas of user needs to represent both the types of knowledge, declarative as well as procedural.

1.4 CONCLUSIONS

Intelligence relates to effective use of rules and facts. A most suitable mechanism is required to adequately represent the facts and rules for a particular domain. Efficient and effective KR is the only way to provide a suitable solution to the issues in AI. A representation is a method to describe the knowledge (both new as well as the one obtained by inferencing the existing knowledge) so that any reasoning system can easily be adopted.

It has been found that many KR systems were developed in different domains and used for different applications. So it made sense to take advantage of technology to serve the people so that they can learn and explore the world. If user wants to represent the reality of the problem domain and KR system is used to solve the problem, then the first check the adequacy of the selected KR concerning with the real world and the problem domain. The adequacy of the chosen representation with respect to the real-world facts and the problems need to be solved. Three types of adequacies for KR system suggested by McCarthy and Hayes (1969) and Sowa (1992) are as *Metaphysical:* Without any contradiction world could be constructed (facts or reality) by KR system. *Epistemological:* The representation system should express the facts practically that resemble the world. *Heuristic:* Reasoning processes should be expressible in the representation, that is, a KR must have all these adequacies. The various KR tools for different applications are presented in Table 1.6.

TABLE 1.6 Knowledge Representation Tools.

Tool Name	Developer	Description	Reference
Rhet (rhetorical) system	James Allen and Brad Miller	Rhetorical is a KR system. It is directed toward a set of tools which build automated reasoning system, with major applications in planning and natural language. It is the extension HORNE KR system. Horn clause, backwards chaining, and a forward chaining were used for reasoning purpose.	James, Allen and Miller, 1991
Shocker	Brad Miller	It is a KR tool that is proposed to build a prototype natural language understanding and planning systems. It is HKR, which consists of dedicated reasoning systems which was presented to the user within a single consistent structure. It was based on Rhet.; it has not been released to the public yet.	Allen and Miller,1992
Concept maps	Mauri Ahlberg	The camp was used to represent educational knowledge graphically. The tree of Porphyry is the earliest known example of tree representation. It was completely based on mind mapping and concept mapping. The camp allows user to build knowledge in both ways individually and collaboratively. It was able to use all digital resources of their own and in WWW, using drag, drop, and clicking to open or to publish at WWW.	Ahlberg,1993
Narrative knowledge representation language (NKRL)	Gian, Piero, Zarri	NKRL is used to represent the semantic content, that is, meaning of complex narrative texts. It has the aspiration of crux informational element of the original narrative expression and has its use in the representation of plural expressions.	Gian, Piero, Zarri,1997
COLAB	Boley, Hanschke, Hinkelmann, and Meyer	It is HKR that emphasis on the horizontal and vertical compilation of knowledge base. Its subsystems deal with different types of knowledge. Each subsystem can act as a standalone system.	Boley et al., 1991

TABLE 1.6 *(Continued)*

Tool Name	Developer	Description	Reference
Classora	Ivan, de Classora	A knowledge base for the Internet oriented to data analysis. It is used to store structured information and displayed them in multiple ways analytically, graphically, geographically. It also used OLAP analysis. Classora-contained information that was collected from public sources.	Ivan, de Classora,
DLOG	Randy Goebel	DLOG is a Prolog-based system. The extended unification procedure was used for reasoning.	Randy Goebel , 1985
LOOM	Robert Macgregor	The LOOM is an application-independent HKR system and is a classification-based KR system. It is a frame-based system and all the statements in LOOM were mapped into the predicate logic, Fikes and Kehler (1985). LOOM provided an intelligent environment for the problem domain. LOOM represents the declarative knowledge consisting of rules, facts, and default rules.	Macgregor, Robert, 1991
A logic-based semantic network processing system (SNePS)	Stuart Shapiro, 1971	SNePS is hybrid of logic, frame, and network KR techniques. Knowledge base (KB) in SNePS has a set of assertions (propositions) about various class/object/entities. SNePS can be used in common sense reasoning, natural language generation and understanding, and for cognitive robotics.	Stuart C.Shapiro, (2000), and Sowa (1992), www.cse.buffalo.edu/~rapaport/676/F01/shapiro.rapaport.95.pdf
YAK	Franconi Enrico	It is based on description logic in which terminological language is a frame-based description language that allows to represent primitive concepts and roles, functional roles and role conjunction. An object-oriented assertion box is used as a recognition reasoning procedure. A constraint box is used to represent disjointness, implies rules, transitive roles, the test operator to handle concrete domains.	Enrico Franconi, 1994

KEYWORDS

- **knowledge representation (KR)**
- **hybrid knowledge representation (HKR)**
- **semantic net**
- **ontology**

REFERENCES

Acar, E. *Computing Subjective Expected Utility using Probabilistic Description Logics, Frontiers in Artificial Intelligence and Applications STAIRS 2014*, pp 21–30. https://ub-madoc.bib.uni-mannheim.de/36878/1/AcarSTAIRS14.pdf.

Ali, A.; Khan, M. A. Selecting Predicate Logic for Knowledge Representation by Comparative Study of Knowledge Representation Schemes. Proceeding of the International Conference on Emerging Technologies (ICET) 2009, Islamabad, Pakistan, Oct, 2009, pp 23–28.

Allen, J. F.; Miller, B. W. *The RHET System: A Sequence of Self-Guided Tutorials,* TR 325, Computer Science Department, University of Rochester, 1991.

Allen, J.; Ferguson, G.; Gildea, D.; Kautz, H.; Schubert, L. Artificial Intelligence, Natural Language Understanding and Knowledge Representation and Reasoning, 2nd ed.; Benjamin Cummings, 1994.

Baader, F.; Calvanese, D.; McGuinness, D. L.; Nardi, D.; Patel-Schneider, P. F. *The Description Logic Handbook: Theory, Implementation, and Applications*, Cambridge University Press, 2003.

Bill, S.; Gil, Y. Flexible Knowledge Acquisition Through Explicit Representation of Knowledge Roles. AAAI Spring Symposium on Acquisition, Learning, and Demonstration: Automating Tasks for Users; Stanford, 1996; pp 138–143.

Bittencourt, G. *An Architecture for Hybrid Knowledge Representation*, Ph.D. Dissertation, Karslruhe University, West Germany, 1991.

Bittencourt, G.; Calmet, J.; Homann, K.; Lulay, A. *MANTRA:* A Multilevel Hybrid Knowledge Representation System, Proceedings of the XI Brazilian Symposium on Artificial Intelligence, Fortaleza, 1994, pp 493–506.

Brachmana, R. J.; Fikes, R. E.; Levesque, J. H. *KRYPTON: Integrating Terminology and Assertion*, 1983. [Online] http://www.aaaipress.org/Papers/AAAI/1983/AAAI83-005.pdf (accessed May 28, 2014).

Brachman, R. J.; Levesque, H. J. 1984. The Tractability of Subsumption in Frame Based Description Languages. In *Proceedings Fifth National Conference on Artificial Intelligence (AAAI-84)*; Morgan Kaufmann Publishers: Austin, 1984.

Brachman, R. J.; Levesque, H. J. *Readings in Knowledge Representation;* Morgan Kaufmann Publishers: San Francisco, 1985.

Brachman, R. J.; Levesque, H. J. 2003. *Knowledge Representation and Reasoning.* http://rair.cogsci.rpi.edu/pai/library/brachmanbook7-17-03.pdf (accessed May 5, 2014).
Brachmana, R. J.; McGuinness, D. L.; Patel-Schneider, P. F.; Borgida, A. Reducing Classic to Practice: Knowledge Representation Theory Meets Reality. *Artif. Intell.* **1999,** *114*(1–2).
Brachmana, R. J.; Schmolze, J. G. *An Overview of the KL-ONE Knowledge Representation System. Cognit. Sci.* **1985,** *9*(2), 171–216.
Brewster, C.; Hara, K.; Fuller, S.; Wilks, Y.; Franconi, E.; Musen, M. A.; Ellman, J., Buckingham Shum, S. *Knowledge Representation with Ontologies: The Present and Future*; IEEE Intelligent Systems, 2004, ISSN 1541-1672.
Buchanan, B. G.; Shortliffe, E. H. *Rule-Based Expert Systems: The MYCIN Experiments of the Stanford Heuristic Programming Project*, Addison-Wesley Publisher, 1984.
Burke, R. R.; Rangaswamy, A., Wind, J.; Elisashberg, J. *A Knowledge Based System for Advertising Design. Mark. Sci.* **1990,** *9*(3), 1990.
Byun, D. H.; Suh, E. H. Assessing key knowledge representation techniques for use in HRM problem domains. *Int. J. Expert Syst. Appl.* **1996,** *11*(3), 287–299.
Calmet, J.; Tjandra, I. A.; Bittencourt, G. 1991b. *MANTRA:* A Shell for Hybrid Knowledge Representation. 3rd International Conference on Knowledge Representation with Ontologies: The Present and Future, Nov 10–13, 1991.
Calmet, J.; Tjandra, I. A. 1994. Building Bridges between Knowledge Representation and Algebraic Specification, Methodologies for Intelligent Systems; *Lecture Notes in Computer Science*, Vol. 869, 1994, [Online] http://www.springerlink.com/content/70051573m2513005 (accessed June 8, 2014).
Christian, R. Object-Oriented Programming and Frame-Based Knowledge Representation, Proceeding of the 1993 IEEE International. Conference on Tools with AI, Boston, Massachusetts, Nov 1993.
Costa, R. Knowledge Representations with Ontology Support for Collaborative Engineering in Architecture Engineering and Construction. *ITcon* **2014,** *19*, 434–461. http://www.itcon.org/2014/26.
Davis, R.; Shrobe, H.; Szolovits, P. *What is a Knowledge Representation* AI Mag. **1993,** *14*(1), 17–33.
Diederich, J.; Roman, I.; Mark, M. KRITON: A Knowledge-Acquisition Tool for Expert System. *Int. J. Man-Mach. Studies* **1987,** 29–40.
Do, N.; Nguye, H. Model for Knowledge Representation using Sample Problems and Designing a Program for Automatically Solving Algebraic Problems, World Academy of Science, Engineering and Technology, 2010.
Esragh, F.; Mamdani, E. H. *A General Approach to Linguistic Approximation in Fuzzy Reasoning and its Applications*; Academic Press, International Publisher: London, 1981.
Fenves, S. J.; Maher, M. L.; Sriram, D.; Anselmo, P. *Information Retrieval Baselines for the ResPubliQA Task.* In *Cross Language Evaluation Forum,* 2009. file:///C:/Users/Birbal/Downloads/bse-pe-003_1985_9_J-29_a_004_d.pdf. (accessed July 14, 2014).
Fikes; Kehler T. The Role of Frame-Based Representation in Reasoning. *Commun. Assoc. Comput. Mach. (ACM)*, **1985,** 28(9), 904–920.
Franconi, E. Description Logics for Natural Language Processing, AAAI Technical Report, 1994, AAAI. www.aaai.org (accessed June 5, 2014).
Fritz, L. *Semantic Networks*; Parsons Avenue, Webster Groves, Missouri, U.S.A.

Frost, R. A. A Method of Facilitating the Interface of Knowledge Base System Components. *Comput. J.* **1985,** *28*, 112–116.

Grimm, S.; Hitzler, P.; Abecker, A. *Knowledge Representation and Ontologies Logic, Ontologies and Semantic Web Languages*, 2007, pp 51–106 [Online]. http://knoesis.wright.edu/pascal/resources/ publications/kr-onto-07.pdf (accessed June 9, 2014).

Hafner, C. D.; Baclawski K.; Futrelle R. P.; Fridmanm, N.; Sampath, S. *Creating a Knowledge Base of Biological Research Papers*; Intelligent System in Molecular Biology ISMB, 1994; pp 147–152.

Hendrix, G. G. *Expanding the Utility of Semantic Networks through Partitioning*; Artificial Intelligence Center, Stanford Research Institute, Menlo Park, California, 1977.

Heflin Douglas, J. Towards the Semantic Web Knowledge Representation in a Dynamic, Distributed Environment, PhD Dissertation, University of Maryland at College Park, USA, 2001.

Henrick, E. 1997. *Specification and Generation of Custom-Tailored Knowledge-Acquisition Tools*. [Online] http://ijcai.org/Past%20Proceedings/IJCAI-93-VOL1/PDF/072.pdf (accessed June 7, 2014).

Hitzler. Presentation on Introduction to Semantic Web. http://www.semantic-web-book.org/w/images/8/87/W2011-01-introduction.pdf (accessed June 7, 2014).

Houben, G. J. P. M. *Knowledge Representation and Reasoning, Dutch Research Database*, 2002.

Jain, S.; Mishra, S. Knowledge Representation with Ontology Proceedings of International Conference on Advances in Computer Engineering and Applications, *ICACEA 2014*, 2014; pp 1–5.

Jeremy, G. *Lecture Notes, Imperial College*, London. [Online] www.doc.ic.ac.uk/~sgc/teaching/ v231/lecture4.ppt (accessed May 28, 2014).

Jeng, S. K. *Lecture Notes on Knowledge Representation*. [Online] www.cc.ee.ntu. edu.tw/~skjeng/Representation.ppt (accessed May 24, 2014).

Johnson, L.; Johnson, N. E. *Knowledge Elicitation Involving Teachback Interviewing in Kidd, Knowledge Acquisition for Expert Systems: A Practical Handbook*, 1987. [Online] http://ksi.cpsc.ucalgary.ca/articles/KBS/KSS (accessed May 18, 2014).

Korb, K. B.; Nicholson, A. E. Chapter on *Introducing Bayesian Networks*, from *Bayesian Artificial Intelligence*, 2nd ed.; 2010. [Online] www.csse.monash.edu.au/bai/book/BAI_Chapter2.pdf.

Kuechler, W. L. Jr.; Lim, N.; Vaishnavi, V. K. A Smart Object Approach to Hybrid Knowledge Representation and Reasoning Strategies, Hawaii, International Conference on System Sciences, HICSS, 1995.

Lecture Notes on Frame Knowledge Representation Technique. [Online] http://userweb.cs.utexas.edu/users/qr/algy/algy-expsys/node6.htm (accessed May 29, 2014).

Lecture Notes on Knowledge Representation Misc Psychology and Languages for Knowledge Representation. [Online] http://misc. thefullwiki.org/Knowledge representation (accessed July 19, 2012).

Lecture Notes on Predicate logic. [Online] http://www.cs.odu.edu/~toida/nerzic/content/logic/pred_logic/inference/infer_intro.html (accessed May 28, 2014).

Lee, T. B. Chapter on Semantic Web Road Map, 1998. [Online] www.w3.org. (accessed May 28, 2014).

MacGregor, R. Using a Description Classifier to Enhance Knowledge Representation, 1991, Source IEEE Explore.

Mahalakshmi, G. S.; Geetha, T. V. Representing Knowledge Effectively using Indian Logic, *TMRF e-Book,* 2009.

Mahalakshmi, G. S.; Geetha, T. V. Gautama Ontology Editor Based on Nyaya Logic and its Applications, Third Indian Conference, ICLA, Chennai, 2009.

Mauri, Å. Concept Maps, Vee Diagrams and Rhetorical Argumentation (RA) Analysis: Three Educational Theory-Based Tools to Facilitate Meaningful Learning, Proceeding of the 3rd International Seminar on Misconceptions in Science and Mathematics, Aug 1–5, 1993. Cornell University. http://www.mlrg.org/proc3abstracts.html (accessed June 5, 2014).

Meyer, J.; Dale, R. Building Hybrid Knowledge Representations from Text, Division of Information and Communication Sciences; Macquarie University, Sydney, NSW, 2109.

Mylopoulos, J. *An Overview of Knowledge Representation*, Department of Computer Science, *University of Toronto*, 1980, ACM0-89791-031-1/80/0600-0005.

Mylopoulos, J.; Wang, H.; Kushniruk, A. KNOWBEL*:* A Hybrid Expert System Building Tool, Proceeding of 2nd IEEE International. Conference on Tools for Artificial Intelligence, 1990.

Okafor, E.; Osuagwu, C. C. Issues in Structuring the Knowledge-Base of Expert Systems. *Electron. J. Knowl. Manag.* **2007,** *5*(3).

Peltason, C. *The BACK System—an Overview, ACM SIGART Bulletin - Special Issue on Implemented Knowledge Representation and Reasoning Systems*. Homepage **1991,** *2*(3).

Nikolopoulos. Expert Systems: Introduction to First and Second Generation and Hybrid, CRC Press, 1997.

Patel-Schneider, P. F. A four-Valued Semantics for Frame-Based Description Languages, Proceeding of American Association for Artificial Intelligence-86, 1986.

Patel-Schneider, P. F. *A Hybrid, Decidable, Logic-Based Knowledge Representation System. Comput. Intell.* **1987,** *3*(1), 64–77.

Presentation on Knowledge Representation and Rule Based Systems. [Online] www.arun555 mahara.files.wordpress.com/2010/02/knowledge-representation.ppt (accessed June 14, 2014).

Presentation on Knowledge Representation Techniques. [Online] http://www.scribd.com/doc/6141974/semantic-networks-standardisation (accessed June 4, 2014).

Presentation on Knowledge Representation using Structured Objects. [Online] www.freshtea.files.wordpress.com/2009/.../5-knowledge-representation.ppt (accessed May 29, 2014).

Presentation on Knowledge representation. [Online] http://www.doc.ic.ac.uk/ ~sgc/teaching/v231/lecture4.ppt (accessed June 5, 2014).

Presentation on Various knowledge Representation Techniques. [Online] http://www.ee.pdx.edu/~mperkows/CLASS_ROBOTICS/FEBR-19/019.representation.ppt (accessed May 14, 2014).

Quesgen, W.; Junker, U.; Voss, A. 1987. *Constraints in Hybrid Knowledge Representation System*, Expert Systems Research Group, F.R.G [Online] http://dli.iiit.ac.in/ijcai/VOL1, IJCAI, 1987 (accessed May 28, 2014).

Rajeswari, P.V.N.; Prasad, T. V. Hybrid Systems for Knowledge Representation in Artificial Intelligence, *Int. J. Adv. Res. Artif. Intell.* **2012,** *1*(8). http://arxiv.org/ftp/arxiv/papers/1211/1211.2736.pdf.

Rathke, C. 1993. Object-Oriented Programming and Frame-Based Knowledge Representation, 5th International Conference,Boston, 1993.

Reichgelt, H. *Knowledge Representation: An AI Perspective*; Ablex, 1991.

Rich, E; Knight, K. *Artificial Intelligence, 2nd ed.*; McGraw-Hill, 1991.

Rodrigo, Á.; Pérez, J.; Penas, A.; Garrido, G.; Araujo, L. Approaching Question Answering by means of Paragraph Validation, In Cross Language Evaluation Forum (CLEF), 2009.

Rosenbloom, P. S. Beyond Generalization as Search: Towards a Unified Framework for the Acquisition of New Knowledge. In *Proceedings of the AAAI Symposium on Explanation-Based Learning*; DeJong, G. F., Ed.; Stanford, CA, 1998; pp 17–21.

Russell, S.; Norvig, P. *Artificial Intelligence: A Modern Approach,* 3rd ed.; Prentice Hall, 2009.

Saffiotti, A.; Sebastiani, F. *Dialogue Modelling in M-KRYPTON, a Hybrid Language for Multiple Believers, Department of Linguistics*; University of Pisa, 1988; pp 56–61.

Schmolze, J. G.; Lipkis, T. A. . Classification in the KL-ONE Knowledge Representation System, Proceedings of the Eighth International Joint Conference on Artificial intelligence (IJCAI'83*)*, 1983; Vol. 1, pp 330–332; source IEEE Xplore.

Sedki, K.; Beaufort, L. Cognitive Maps and Bayesian Networks for Knowledge Representation and Reasoning, 24th International Conference on Tools with Artificial Intelligence, Greece, Aug 2012.

Shapiro. Representing and Locating Deduction Rules in a Semantic Network, SIGART Newsletter No. 63, June 1977, pp 14–80.

Sharif, A. M. Knowledge Representation within Information Systems in Manufacturing Environments, Brunel University Research Archive, 2004.

Sharma, T.; Kelkar, D. A Tour towards Knowledge Representation Techniques. *Int. J. Comput. Technol. Electron. Eng.* **2012,** *2*(2), ISSN 2249-6343.

Shetty, R. T. N.; Riccio, P. M.; Quinqueton, J. 2006. Hybrid Model for Knowledge Representation, ICHIT '06 Proceeding of the Inernational Conference on Hybrid Information Technology, 2006; Vol. 1, pp 355–361.

Shetty, R. T.N.; Riccio, P. M.; Quinqueton, J. Extended Semantic, Network For Knowledge Representation, Information Reuse and Integration, France, 1987, IEEE-IRI, 2009; Vol. 1.

Siddique, N. H.; Balasundram, P. A.; Ikuta, A. 2008. Editorial: *Hybrid Techniques in AI*, Springer Science+Business Media B.V., 2008.

Singhe, R. A., Rana C.; Ajilh, P. M. A. Enhanced Frame-based Knowledge Representation for an Intelligent Environment, IEEE, KIMAS, Boston, USA, 2003.

Sowa, J. F. *Encyclopedia of Artificial Intelligence*, 2nd ed.; Wiley, 1992.

Stillings and Luger. *Knowledge Representatio*n, Chapters 4 and 5, 1994. [Online] http://www.acm.org/crossroads/ and www.hbcse.tifr.res.in/jrmcont/ notespart1/node28.htm (accessed May 18, 2014).

Shapiro, S. C. SNePS: *A Logic for Natural Language Understanding and Commonsense Reasoning*, AAAI Press/MIT Press, 2000; pp 175–195.

Russell, S.; Norvig, P. *Artificial Intelligence: A Modern Approach,* 3rd ed.; Prentice Hall, 2009.

Studer, R.; Grimm, S.; Abecker, A. *Ebook on Semantic Web Services*: Concepts, Technologies, and Applications [Online] http://www.racer-systems.com (accessed May 18, 2014).

Szolovits, P. 1977. *An Overview of OWL, A Language for Knowledge Representation,* Presented at the Workshop on Natural Language for Interaction with Databases held by the

Institute for Applied Systems Analysis (IIASA) at Schools Laxenburg, Austria, in January, 1977. [Online] http://groups.csail.mit.edu/medg/people/psz/ftp/OWL_overview1.html (accessed June 9, 2014).

Tanwar, P.; Prasad, T. V. 2010. Comparative Study of Three Declarative Knowledge Representation Techniques. *Int. Comput. Sci. Eng.* **2010**.

Tanwar, P.; Kaur, M.; Prasad, T. V. Knowledge Representation using Semantic Net and Fuzzy Logic, Nat. Conference on Advances in Knowledge Management, Lingaya's University, Faridabad, India, 2009.

Tanwar, P.; Prasad, T. V.; Dutta, K. An Effective Knowledge Base System Architecture and Issues in Representation Techniques. *Int. J. Adv. Technol.* [Online] http://ijict.org, ISSN 0976-4860, 2011; Vol. 2, pp 430-437.

Tanwar, P.; Prasad, T. V.; Dutta, K. 2012. Hybrid Technique for Effective Knowledge Representation & a Comparative Study. *Int. J. Comput. Sci. Eng. Surv.* **2012,** *3*(4), pp 43–57.

Tanwar, P.; Prasad, T. V.; Dutta, K. Hybrid Technique for Effective Knowledge Representation. *Springer Book Series Title Adv. Intell.t Syst. Comput.* **2012,** *178*, 33–43.

Tanwar, P.; Prasad, T. V.; Datta, K. Hybrid Technique for Effective KnowledgeRepresentation. In *Advances in Computing and Information Technology. Advances in Intelligent Systems and Computing*; Meghanathan, N., Nagamalai, D., Chaki, N., Eds.; Springer, Berlin, Heidelberg, 2013; Vol. 178.

Tanwar, P.; Prasad, T. V.; Dutta, K. "An Effective Reasoning Algorithm for Question Answering System". *Int. J. Adv. Computer Sc. Appl.* Special Issue on Natural Language Processing, ISSN 2156-5570 (Online), ISSN 2158-107X (Print), April 2014, pp 52–57.

Tanwar P.; Prasad T. V.; Datta K. "A Tour Towards the Various Knowledge Representation Techniques for Cognitive Hybrid Sentence Modeling and Analyzer." *IJICT* **2018,** *7*(3) http://doi.org/10.11591/ijict.v7i3.pp124–134.

Tanwar, P.; Prasad T. V.; Datta, K. Hybrid Technique for Effective Knowledge Representation. In *Advances in Computing and Information Technology. Advances in Intelligent Systems and Computing*; Meghanathan, N., Nagamalai, D., Chaki, N., Eds.; Springer: Berlin, Heidelberg, 2013, Vol. 178.

Thomas, F. Gordon, Oblog-2 A Hybrid Knowledge Representation System for Defeasible Reasoning, ACM O-89791, pp 231–239.

Tian, Y.; Wang, Y.; Gavrilova, M. L.; Guenther Ruhe. A Formal Knowledge Representation System for the Cognitive Learning Engine, Proceeding of the 10th IEEE International Conference of Cognitive Informative and Cognitive Computing, 2011.

Tsai, J. J. P., Aoyama, M.; Change, Y. L.. Rapid Prototyping using FRORL Language, Department of Electrical Engineering and Computer Science, University of Illinois at Chicago, Chicago, 1998, source IEEE Xplore.

Tsai, J. J. P.; Jang, H. C.; Schellinger, K. J.. RT-FRORL: A Formal Requirements Specification Language for Specifying Real-Time Systems, Dept. of Electrical Engg. and Computer Science, University of Illinois, Chicago, 1991, source IEEE Xplore.

Tsai, J. J. P.; Weigert, T.; Jang, H. C.,. A Hybrid Knowledge Representation as a Basis of Requirement Specification and Specification Analysis. *IEEE Trans. Softw. Eng.* **1992,** *18*(12).

Varga, E.; Jerinic, L. Knowledge Representation With Semantics Network Of Frames Technique, Proceedings of the 38th Conference of Electronics, Telecommunications, Computers, Automation and Nuclear Engineering, Belgrade, June 1994, pp 217–218.

Wai, K. Semantic Modeling and Knowledge Representation in Multimedia. [Online] ieeeexlore.ieee.org, 1999 (accessed May 28, 2014).
Web Document on Predicate Logic, History. [Online] http://www.cs.bham.ac.uk/research/projects/poplog/computers-and-thought/chap6/ node5.html (accessed June 9, 2014).
Yue, M.; Zuoquan, L. Inferring with Inconsistent OWL DL Ontology: A Multi-Valued Approach. Proceedings of the International Conference on Semantics in a Networked World, ICSNW-2006, Munich, Germany, Springer-Verlag, 2006.
Zadeh, L. A. Computing with Words using Fuzzy Logic, *IEEE Transactions*, 1993.
Zhang, N. L.; Guo, H. *Chapter on Introduction to Bayesian Networks*; Science Press: Beijing, 2006.

CHAPTER 2

Understanding Distributed Semantic Analysis with Spark Data Frames

RICHA MATHUR[1*], DEVESH K. BANDIL[2], and
DHANESH KUMAR SOLANKI[3]

[1] *C 404, Aarshiwad, Anandam, Balaji Market, Shrinathpuram, Kota, Rajasthan*

[2] *Suresh Gyan Vihar University, Jagatpura, Jaipur*

[3] *E 325 III-B Khetri Nagar, Jhunjhunu, Rajasthan*

[*] *Corresponding author. E-mail: richa0058@gmail.com*

ABSTRACT

Big data analytics is the most current research trends till date. Big data faces many challenges in terms of processing and storing of data for optimization and implementation of machine learning (ML) algorithms as ML needs too much computation power to process data for any kind of output in terms of analytics. To reduce the optimization problem, we are using the spark for its data frames. Also, in creating the architecture of any big data problem, spark engine provides very good support in that too as its engine support any archistecture with Hadoop, map-reduce. Analytics in big data need real-time processing for any behaviors that we needed to predict under any type of text analytics.

While working on semantic analysis, we have faced many problems as there are many different spelling mistakes in documents from online sources. To handle this problem, we are creating different distributed schema under different layers of distributed data frames of spark engine with support of YARN, as this Yarn supports us for any kind of map-reduce process we need to follow to solve sematic behaviors of any mistakes that happened inside sources. But, the first step that we followed is to create ontology from the above semantic characteristics. As the base architecture of the ontology helps us to understand how data behave under different processes of semantic field under any language used in many datasets while working with Natural Language Processing (NLP), we need to take care of the sentiment of the document to parse the information under positive, negative, or neutral in term of sentiment analysis.

In this research, we mainly focus on ontology architecture to process different semantic information to increase the sentiment accuracy by implementing different advance ML.

2.1 INTRODUCTION

The field of big data analytics is becoming very popular these days because organizations across the world adopt their ability to leverage insights to gain competitive advantage. As big data analytics are useful in making better decisions and draw certain conclusion on specific topic. Some elements of big data analytics include predictive models, statistical algorithms, and high-performance analytic system that can perform complex computation. Big data include various technologies like Hadoop, Spark, MapReduce, etc.

Semantics (a linguistic term) is the study of meaning of words in a language or logic. It focuses on the study of relation between signifiers, like words, phrases and symbols, and their denotation. Style of any language plays an important role in understanding the meaning of code. Semantics tries to understand about the language's construction, interpretation, clarification, illustration, contradiction, and negotiation by its speakers and listeners. Semantic information is useful for all told aspects of understanding natural language. Natural language includes expounded linguistics analysis with the structures and occurrences of the words, phrases, clauses, paragraphs, etc., and perceives the thought of what's communication or logic. Extracting information from semantics is difficult to get; however, it will add power and accuracy to natural language processing (NLP) systems.

Web-based social networking destinations become well known now daily as individuals over the world relies on them to speak with their relatives, companions, and rest of the world. Sentiment analysis (SA), otherwise called opinion mining, is the most mainstream pattern in these days and ages which is the way toward recognizing and sorting conclusions on the web, decides the author's frame of mind toward a theme or item. It advises about what creator needs to convey and characterizes his perspective regarding feelings, emotions, and subjectivities about an occasion or theme.

Ontology is a formal portrayal of classes, properties, and relations between ideas, information and substances inside an area. Ontologies are made to confine multifaceted nature and arrange data into data and knowledge.[7] Meaning of ontology can be shaped with the assistance of sharing regular comprehension of learning among individuals and programming specialists that makes domain information reusable. The ontology model could be put away in OWL/XML or RDF/XML group. We can recover data sorted out in an ontology model by questioning it.[1]

Finding significant data among the gigantic measure of information on the Web is a troublesome undertaking. In Semantic Web, ontologies are utilized to appoint (concurred) which means to the substance of the Web. So in this paper, we are focusing on ontology architecture to process semantic information by implementing different machine learning algorithms to increase sentiment accuracy.

2.2 RELATED WORK

Zhao et al.[8] explored the effects of various preprocessing methods for classifying sentiments of twitter datasets. They presented some preprocessing methods for removing unwanted data such as removal of URLs, numbers, stop words, replacing negations, reverting repeated words, and expanding acronym. Two feature models, namely word n-gram and prior polarity score, are used on five twitter datasets to identify the polarity of tweet sentiment. Four popular supervised classifiers are used to analyze the effect of preprocessing such as Naive Bayes, SVM, Logistic Regression, and Random Forest. Their experimental result shows that by removing stop words, numbers, and URLs from sentiments noise level can be reduced but it cannot improve performance much more. By replacing negations and expanding acronym, efficient results can be obtained and it also can improve the accuracy of classification methods.

Krouska et al.[4] described the role of preprocessing techniques for classification methods. They used three-different sub-datasets to investigate the performance of some ML classification-based algorithms with different preprocessing options. For preprocessing purpose, they presented TF-IDF weighting scheme, stemming, stop-word removal, and Tokenization technique. To increase the classification of accuracy and reduce training time, they used feature extraction model. For classifications of tweets, four machine learning (ML) calculations are utilized by them in particular Naive Bayes, SVM, K-Nearest Neighbor, and Decision Tree. Their test result demonstrates that suitable feature determination and representations can improve execution for different characterization techniques.

2.3 PROBLEM STATEMENTS

As users on social media sites are rapidly growing and producing a large amount of data every day, so there is a need to store, process, and analyze these messages to find out its polarity on some topics or events. We have faced many problems as there are many different spelling mistakes in documents while working on semantic analysis. To understand the semantic behaviors, we need to analyze sentiment of the document to parse the information under positive, negative, or neutral.

Optimization and implementation of correct ML algorithm and extracting appropriate results (in the form of sentiments), that is, based on queries after processing semantic data, are challenging task.

2.4 BIG DATA ANALYTICS WITH SPARK

Big data is a wide term for any voluminous and compound datasets that can provide useful information after processing. The information collected is not in suitable form for analysis. We require an information extraction process, that is, able to find out the required information from unstructured or raw data and put it in structured format. Data cleaning assumes predefined constraints to check valid data or error models for some domains. Data analysis helps to understand relationship among objects, develop methods for data mining to predict future observations accurately, enabling interactive response time but has lack of coordination between databases systems that is the current problem with big data. Better basic leadership and vital planning in business should be possible by examining big data. Enormous data examination is the way toward dissecting data collections and concentrate results from it to increase better bits of insights and opportunities. Exponential development of data in all fields requires total measures for getting to and overseeing such sort of information. Enormous data are for the most part connected with distributed computing, as investigation of huge datasets requires Hadoop like platform to store huge datasets over a distributed cluster and MapReduce to arrange, consolidate, and process information from numerous sources.

Hadoop gives an open-source framework for storing data in distributed manner and processing applications on exceptionally enormous datasets. Map reduce framework is the primary part in Hadoop system that is utilized for processing and creating huge datasets on a cluster with distributed or parallel algorithm.

Where, Spark is a framework for performing general data analytics on Hadoop like distributed computing cluster by providing in-memory computation for process data and increase speed over Map Reduce. It is a fast and general engine (supports multiple language faster than Hadoop) for large scale data processing that runs on the top of Hadoop cluster. Features of Spark include in-memory computation, real-time stream processing, and advanced analytics (with SQL queries, ML algorithms, Graph algorithms, and streaming data). Spark works like a "library" that enables parallel computations via function calls using MLlib API. The main feature of Spark is Resilient Distributed Dataset (RDD), which stores data in-memory in a fault tolerant (helps in recovering from failures) and parallel way.

2.5 SEMANTIC ANALYSIS

Semantic analysis expects guidelines to be characterized for the system that are same as the manner in which we consider a language and we request that the PC mimic. For instance, "Car is white" is a straightforward sentence in which a human comprehends that there is something many refer to as Car and it is white in shading and the human realizes that red methods shading. Semantic examination likewise incorporates disambiguation of words with different faculties. Representation of semantic incorporates just one feeling of chosen words, for example allowed by semantic disambiguation. For instance, "Apple" can mean either as a natural product or an organization name. Computers do not see such sort of language. For that, the expression "Semantic" utilized here, which demonstrates that sentence arrangement has a structure.

Feedback of semantic plays an important role for the development of organizations as semantic analysis helps them draw appropriate results. We center on the recurrence of the words, syntactic structure of the natural language and other language-related components in semantic search. Here, the system comprehends the definite prerequisite of the search question. When we look for "Apple" in Google, it restores the most fitting reports and site pages in regards to the renowned organization in spite of acclaimed fruit with a similar name since the internet searcher comprehends that we are scanning for an organization. Be that as it may, there is huge number of renowned things on the planet and it is very difficult to store all the data physically and recover precise outcomes when asked by a user.

Data can be deciphered with the help of semantic technologies by recognizing the corresponding context when making and representing to complex data and relations between concepts. With semantic technologies, it ends up more obvious the importance and motivation behind data (e.g., images, words, and so forth.) and complex ideas, just as offer learning for people and machines. Semantic technologies can help the more proficiently discovering data since it depends on metadata that contain more data about other data. Standard guidelines are essential for representation to metadata in a formal manner by which metadata can be traded between data frameworks, applications, and workstations. Consequently, for portraying metadata in web setting, the resource description framework (RDF) can be utilized. Making and deciphering results for information systems, semantic technologies can be founded on straightforward methodologies, for example, glossaries (arrangements of words and their definitions), taxonomy or scientific categorizations (pecking orders for terms), and thesauri (relations of similitude and equivalent words) to maintain a strategic distance from grammatical and semantic issues.[5] These Semantic methods work to store data and bring results while questioning.

2.6 SENTIMENT ANALYSIS

Opinions or sentiments of huge people around the world can perform analysis and future predictions. The procedure of SA is identifying the logical polarity of content as far as positive, negative, or neutral. Organizations across the world widely adopted the ability to extract insights from these sentiments of various social media sites. It helps organizations to make predictions of a certain product, reviews, and other decision-making processes that will ultimately increase the profit. So, ultimately SA is beneficial for organizations and individuals to improve their profit as per user or market demand.[8]

The association between the PCs and the human/natural language is called asNLP.[6] NLP technique facilitates easy pre-processing of text, that is, NLP cleans and normalizes text for sentiment analysis. Analysis of sentiments can be based on single phrase or sentence, where the sentiment of the whole sentence is calculated.

2.7 ONTOLOGY

In the field of AI, ontology is a formal unequivocal depiction of ideas in an area of classes, where properties of every idea describe different features, and relation between these ideas.

Ontologies depend on models that will be clarified, arranged by expanding level of semantic wealth, while none of the models accomplishes the level of semantic wealth that ontologies give. For instance, a glossary is a rundown of words in order request with meanings of words, however, with no clarification of relations between these words.[5]

2.8 ONTOLOGY ARCHITECTURE

In OWL or Ontology file, we characterize different OWL classes and relations between them. When SPARQL querying is done on RDF's utilizing ontology file, these relations characterized prove to be useful to bring most fitting outcomes.

Spark has three different data structures available for processing data through its APIs: RDD, data frame, and dataset. Data frame is much faster than RDD because it has metadata associated with it, which allows Spark to optimize query plan. A distributed collection of data called as Spark data frame which is sorted out into columns that give tasks to filter, grouping, or compute totals, and can be utilized with Spark SQL. With the help of structured data files, existing RDDs, tables in Hive, or external databases, these data frames can be developed.

1. Unstructured Text Data Storage in RDF Form

Tools like "Protégé" are accessible to make ontologies. In light of the ideas, ontologies are made with the goal that we can discover proper outcome in a search engine. Unstructured form of available text report requires a semantic structure.

A framework called as RDF is used to store the information given in text into triples form. To concentrate required information, ML and text analytics are utilized with the goal that information can be store in type of triples. The user can look for the required data by questioning finally. To make the search engine attainable to look in different big data parts are important. Ontology and structured type of text data both are put away as RDF's triple.

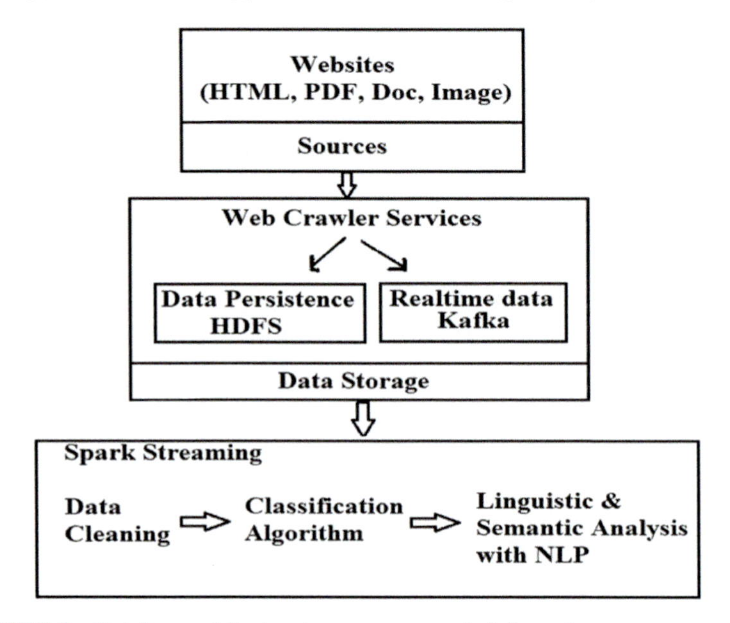

FIGURE 2.1 Ontology architecture to process semantic information.

2. Spark Streaming Used to Process Real-Time Data

In our ontology design, semantic data processing layer uses Spark Streaming to get constant real-time events from Kafka. If there should be an occurrence of "out of memory" issue, we will have a postponement in Spark work, it is seen that Garbage collection delays the Spark work. In such cases, the G1GC calculation can be utilized that aide in tuning runtime parameters of the garbage collector.

After this progression, we process the data for cleaning and pre-preparing. Web-induction techniques are utilized to clean HTML pages and various strategies are utilized for various record types.

To get accurate results, we must make sure that these datasets processed efficiently by removing unrelated contents and thus related contents are accurately extracted. As most researchers consider that URL does not have any information regarding sentiments, so by removing short URLs from datasets, contents can be refined. People often use emotional words that contain repeated letters to express their sentiments which are very common trends like "coooool." Also, numbers are not used for analyzing sentiments so contents can be refined by removing them. The polarity of the word will be changed when they are preceded by a negation or negation can change/reverse the meaning of words. By checking negations, removing of URLs, emotions, numbers, and repeated word; noise in datasets can be reduced.

WEKA tool performs preprocessing, for this we use "StringToWord-Vector" filter that is utilized to change over string attributes into a set of attribute representing word occurrences information from the content contained in the strings. This filter provides us options to do configuration with our dataset that includes the following steps:

- **Stemming:** It is utilized to expel suffix from the word as per some linguistic principles. Here, we apply most prevalent Snowball stemming library.
- **Stop word extractor:** Some words, which do not have polarity, so they do not need to be further analyzed like: able, are, both, which, has, become, after, etc. So, after elimination of these words, our result will not be affected. We used Rainbow list for our experiment.
- **Tokenization:** It is used to split a document into words or terms and makes a word vector. Here, we used NGramTokenizer.
- **Feature selection:** The quantity of attributes is reduced with this method into an improved subset. It can increase accuracy additionally, it gets a decrease training time. Filters and Wrappers are utilized to finish this task. WEKA tools "AttributeSelection" filter is used to select an attribute evaluation method. Here, we use "cfsSubsetEval" method that considers about the individual predictive capacity of each element to assess the value of an attribute.

Preprocessed selected data are saved to HDFS storage framework where replication can be utilized to maintain a failure of node.

3. Machine Learning Algorithms for Data Extraction

Classification and pattern analysis are studied in the field of AI called as ML that enables computer to take in practices of data taken from datasets and makes accurate decisions dependent on data. Here, the steps for extracting information from the data using machine learning techniques are as follows:

- A corpus is created and certain set of rules on this corpus are initiated to classify our dataset.

- The classification algorithm is used to classify the extracted data to find out if it belongs to required area or not. Supervised learning techniques are used for this purpose.
- To run the supervised model, we require training data. Here, training data are manually collected from various sources. The training data finally contain the belonging area.
- This training data are used to run on the supervised model.

Different procedures can be utilized to separate data and all the removed data are stored in RDF form structure.

4. Linguistic Analysis of Text

Linguistic analysis is utilized for data extraction in which standards are defined to separating feature from our text data. Different annotations like person, organization, location, company and so forth are utilized to remove significant features from content or text data.

At that point, we have to recognize the relation between sentences. Semantic analysis enables us to manufacture relations between text data and further they are converted over into triples for storage.

Triples mainly contain three components:

- **Subject**—The URL, document, the individual are represented by the Subject.
- **Predicate**—Stores relations.
- **Object**—Relation values hold by objects for the specific subject.[7]

These triples are then stored in Triple DataStore (RDF).

5. Query Expansion Automatically with Linguistic Perspective

Goal of Semantic Search Engine is to comprehend what precisely user needs. With the assistance of keywords, client can look through their requirements. The system automatically recognizes ontology relations from user search query utilizing linguistic and semantic analysis.

At first, the user query is processed for text analysis purpose to identify the words related with the domain. These words are additionally extended utilizing the ontology file. After that, relations are utilized to extended words that will get scores.[7]

6. Query Ranking Algorithm

To index RDF data on the basis of domain, Apache Solr can be used. Text data are converted into vector form with the help of TD-IF algorithm. For comparability score, cosine likeness is estimated on the extended question (on ontology file) and RDF data. So with this approach ranks are assigned to documents based on their scores.[7]

2.9 CONCLUSION

The procedure of semantic analysis centers on understanding natural language such that human communicate, and as per that getting machine learn. Likewise, it changes over the unstructured text data into structured RDF triples significantly increases with the assistance of text analysis, ML techniques, semantic, linguistics, and phonetic procedures. These RDF triples can be put away in database, and with the utilization of SPARQL question on RDF's triple to extricate the resultant documents in the ranking order dependent on the comparability among query and documents. Feedback is significant for the organizations and its criticism assessment assumes a significant job in the advancement of organizations. The feedback assessment framework comprises of the semantic analysis for deciding the entity of an organization. We have utilized ML algorithms to investigate complex patterns and classification purposes that make on insightful choices dependent on data. Utilizing linguistics and semantic strategies for document recovery aides are accomplishing close flawless AI search engine for any domain.

KEYWORDS

- **machine learning**
- **big data**
- **spark**
- **semantic analysis**
- **ontology architecture**
- **text mining**
- **sentiment analysis**

REFERENCES

1. Thakor, P.; Sasi, S. In *Ontology-Based Sentiment Analysis Process for Social Media Content*, INNS Conference on Big Data, Procedia Computer Science, *53*, 199–207, 2015.
2. Wakade, S. et al. Text Mining for Sentiment Analysis of Twitter Data. The University of Akron, Department of Computer Science.
3. Hemalatha et al. Sentiment Analysis Tool using Machine Learning Algorithms. *Int. J. Emerg. Trends Tech. Comput. Sci.* **2013,** *2(2).*
4. Krouska, A. et al. The Effect of Preprocessing Techniques on Twitter SA. DOI: 10.1109/ IISA.2016.7785373, 7th International Conference on Information, Intelligence, Systems & Applications (IISA), ResearchGate Conference: 2016
5. Eine, B.; Jurisch, M.; Quint, W. Ontology-Based Big Data Management. 2017, https://www.mdpi.com/2079-8954/5/3/45/pdf
6. https:/en.wikipedia.org/wiki/Ontology_(information_science)
7. https://www.xenonstack.com/blog/data-science/semantic-search-machine-learning-nlp-ai/
8. Jianqiang, Z.; Xiaolin, G. Comparison Research on Text Pre-processing Methods on Twitter Sentiment Analysis. *IEEE* Access DOI 10.1109/ACCESS, 2017, 2672677
9. Vinodhini, G. et al. Sentiment Analysis and Opinion Mining: A Survey. *Int. J. Adv. Res. Comput. Sci. Soft. Eng.* 2012, *2*(6).
10. Anber, H. et al. A Literature Review on Twitter Data Analysis. *Int. J. Comput. Electr. Eng.* 2016.
11. Mulay, S. A. et al. Sentiment Analysis and Opinion Mining with Social Networking for Predicting Box Office Collection of Movie. *Int. J. Emerg. Res.*

CHAPTER 3

Machine Learning and Artificial Intelligence in Healthcare

JEELANI AHMED* and MUQEEM AHMED

Department of CS and IT, Maulana Azad National Urdu University, Hyderabad, India

**Corresponding author. E-mail: jeelani.jk@gmail.com*

ABSTRACT

Machine learning (ML) is the application of artificial intelligence (AI) that enables computers to learn automatically, to make improvement by experiencing different scenarios without being explicitly programmed. ML works based on its primary objective that allows computers to learn freely and independently without any kind of human assistance or intervention and adjusts its functions accordingly. The general idea behind ML is that a computer is provided by a training set of examples based on which it performs the task, learns by the experience, and performs the same task with data it has not encountered before. ML helps in analyzing the gigantic volume of information; it generally gives accurate and quick results and may require some extra time and resources to train it properly. AI is revolutionizing health-care industry by affecting many cases in more positive, substantial ways. AI is supplementing doctors, hospitals, and insurance companies. The various stakeholders of health-care sector around the globe are looking for patient-centered, technology-enabled smart health care both inside and outside hospital walls by developing cost-effective and innovative methods. This chapter discusses the emergence and applications of AI in health care by outlining different ML trends in health-care sector.

3.1 INTRODUCTION

The health-care industry is growing at a very fast rate. The spending in global health care is increasingly day by day and is projected to be 4.1% yearly. There are several ways in which artificial intelligence (AI) is being utilized in health-care sector like (1) management of medical records, (2) treatment design, (3) virtual nurses, (4) medication management, (5) drug creation, (6) health monitoring, and (7) performing repetitive jobs.[2] The AI has potential that extends the thinking power of humans in three critical areas: advanced computations, statistical analysis, and hypothesis generation. Health-care providers have started to explore applications of AI algorithms in many areas like skin cancer diagnostic, analyzing lab results, insurance verification, and medical image processing.[3] According to experts, integrating technology with diagnostics can improve precision making and value-based care along with diagnostics itself. AI has an ability to perform millions of complex operations and correlation-finding calculation in a short period of time: the advantage is taken by the human diagnostic team to quickly identify patterns and associations that they would otherwise miss.[4] Combining AI with diagnostics, doctors' responsibility becomes to maintain relationship with patients and AI looks patient's health record and suggests the prescription by matching the symptoms and its cure throughout the world. It is said that AI has less bias and a better memory.[5]

3.2 MACHINE LEARNING STRATEGIES

Computers are made possible to learn automatically and make improvements by experiencing different scenarios without being explicitly programmed by using machine learning (ML) techniques. The main purpose of this technology is to enable computers to learn automatically without any human support.

The general idea overdue this is that a computer is provided by a training set of examples based on which it performs the task and learn by the experience and do the same work with new data.[6] Huge amount of data can be analyzed with the help of ML, and it also gives fast and truthful outcomes. It may need extra while and assets to train it correctly.

The ML algorithms are of the following types:

a. Supervised Machine Learning Algorithms

In this type of learning, the training set of data contains the data and correct outcome of the task with the same data. These leaning algorithms use the past leaning experiences and predict the future.

It produces an inferred function to predict the future output and even compare its output with the correct output, or intended output and find errors. Supervised learning is like giving problems and its solutions to a student and asking him or her to describe a way to solve the future problems.

Classification and regression are the algorithms, which belong to the supervised learning. Classification algorithms receipt input data set alongside the class of each bit of information with the goal that a system figures out how to organize the information. Classification algorithms make usage of support vector machines, classification trees, artificial neural networks (ANN), random forests, and logic regression.

Attribute's value can be predicted by regression algorithms. Regression algorithms include linear regression, Fuzzy classifications, Bayesian networks, decision trees, and ANN.

b. Unsupervised Machine Learning Algorithms

Another type of ML algorithm is the unsupervised ML algorithm in which it is the responsibility of the system to provide the solution on by itself because in training set only data are present not the solution. These algorithms are used when the training set is not classified. This learning discovers the hidden structure of the unlabeled data and infers a function to describe the nature of the data set. This is like asking a student to discover the fundamental reasons that produced the given pattern.

Clustering and dimension reduction are the part of unsupervised ML. Clustering takes input data set and divided it into various groups fulfilling some conditions. Various clustering algorithms like genetic algorithms, Gaussian models, hierarchical clustering, *K*-means clustering, and ANNs are employed.

Dimensional reduction algorithms takes the data set comprises various dimensions and reduces it to lesser dimensions and then fundamental aspects of the data can be captured by these fewer dimensions.

Multidimensional statistics, tensor decomposition, principal component analysis, random projection, and ANNs are some of the dimensional reduction algorithms.

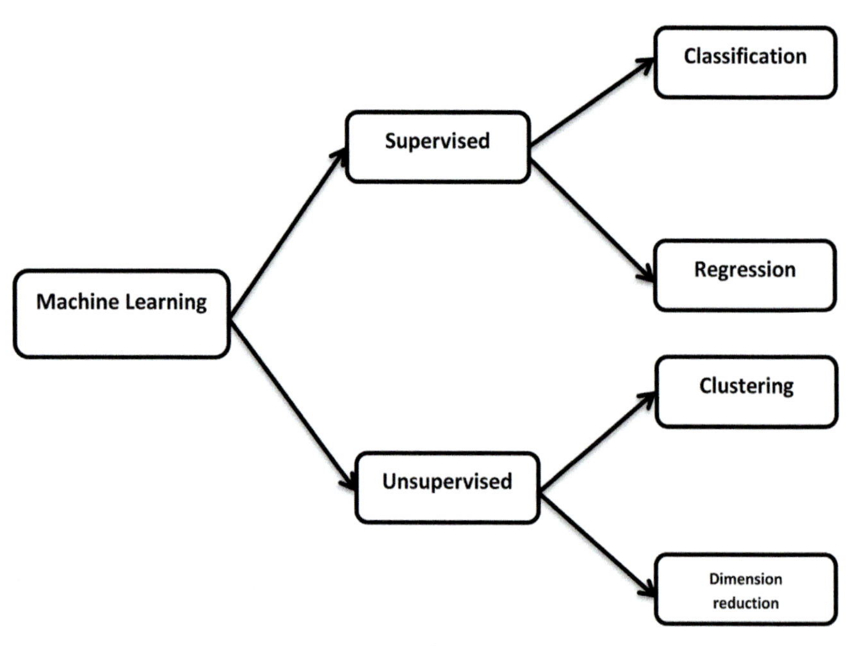

FIGURE 3.1 Machine learning approaches.[6]

c. Semi-supervised Machine Learning Algorithms

This learning falls in between supervised and unsupervised learning. Here, both labeled and unlabeled data set are used to train computers. The input data set contains a lesser quantity of labeled data and big quantity of unlabeled data. The computer system that uses this method is able to improve their learning accuracy.

d. Reinforcement Machine Learning Algorithms

In this method of ML, the system is allowed to interact with the given environment as it produces actions and automatically learns by trial and error. A simple feedback mechanism is required for the system to learn which action is best. In reinforcement ML algorithms, computers learn the ideal behavior in a particular context and it results in maximization of performance.

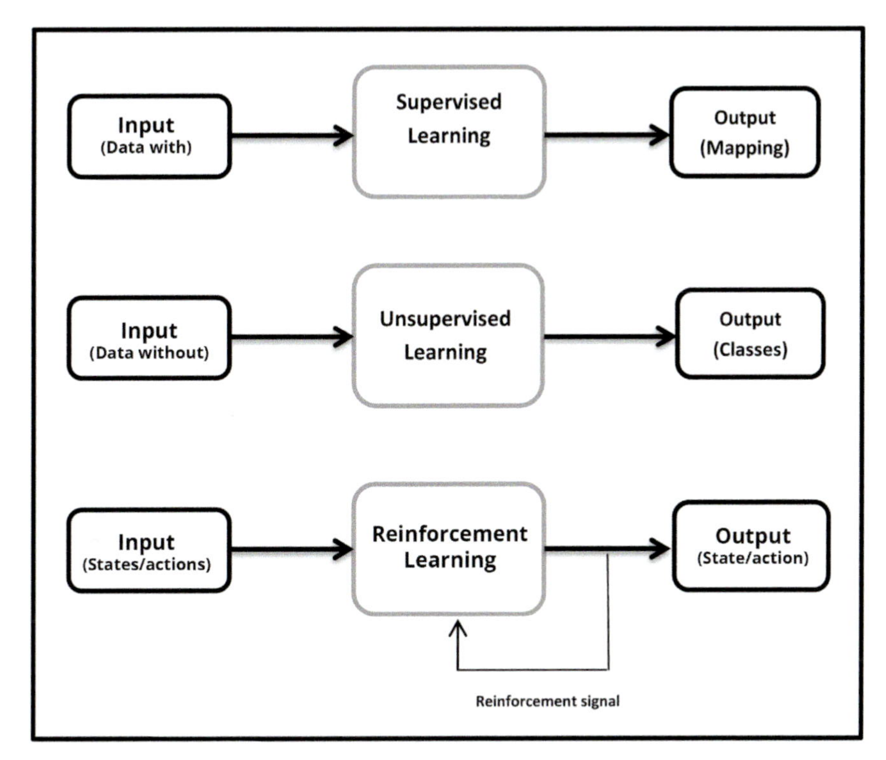

FIGURE 3.2 Illustration of unsupervised learning, supervised learning, and semisupervised learning. X, input dataset; Y, correct outcome.[7]

3.3 HEALTH-CARE DATA

Before deployment of AI systems in health-care applications, these systems must be taught by input data which are produced from various medical deeds, like diagnosis, screening, treatment task, etc. with the goal that system can gain by different associations between subject features, subjects of similar groups, and outcomes of interest.[7] The clinical data exist in various data forms, structured, unstructured, and semistructured, like medical notes, electronic recordings, demographs, physical examinations, images, and data generating from clinical laboratory. The structured and unstructured health-care data can be analyzed by the following components in AI.

The structured data like imaging, genetic, and electrophysiological (EP) data can be analyzed directly by using ML techniques. The ML methods endeavor to group patient's attributes, or gather the plausible disease results.[8]

The natural language processing (NLP) technique enhancements ML to extricate data from unorganized information, for example, clinical notes and medical diaries that advance organized medical information. The NLP method converts unstructured data to machine understandable electronic medical record (EMR)[9], which can be further analyzed by ML methods.

3.4 MEDICAL ARTIFICIAL INTELLIGENCE

Recently, AI methods have sent strong signals across health care; the discussion has started whether AI doctors will slowly replace human physicians in near future, this is due to the vast applications of AI in obtaining better clinical decisions and AI started replacing human judgment in certain fields of medical like in radiology. The rapid development of health-care data which are increasing day by day has made possible the successful applications of AI in health care. The powerful techniques are unlocking hidden patterns and clinically applicable data in the monstrous measure of health-care information that results in assisting clinical decision making.[9–11] The various gains of AI are broadly available in the medical works.[11–13]

From the vast health-care data, AI learn to identify patterns/features using sophisticated algorithms and the obtained AI insights are used in assisting clinical practices. The AI has learning and self-correcting abilities based on the feedback and ultimately it results in maximizing accuracy and performance. AI assists physicians by providing medical information from journals, textbooks, and clinical practices worldwide.[14] In addition, the therapeutic and diagnostic errors can be reduced with the help of AI that are inevitable in the human clinical practice.[9, 10, 14–18]

The health-related outcome predictions and health risk alerts can be done with the help of AI as the system has an ability to extract and infer the useful information from large patient population across world[19].

ML is diminishing the increasing expense of health care and assisting doctors to maintain good relationship with patients in a manner to decide progressively customized medicines and medications and also helping patients in scheduling follow-up appointments.[20]

3.5 AI DEVICES AND RESEARCH INITIATIVE

To advance the lives of patients over the world, there are many research activities that intend to apply ML methods in the health-care sector.[21]

AI is providing expert advice which in result helping more than 30 million Americans who are suffering from diabetes. Glooko's mobile app that runs on iPhone and android-based smartphones helps patients with diabetes to control it by using data around how food, exercise and insulin are affecting their glucose levels under different conditions. Glooko is developing advanced AI algorithms that make use of GPS data from smartphones. With the help of these GPS data, the Glooko might be able to identify the patient that has entered a restaurant and it automatically examines the menu of a restaurant and suggests three best meals for diabetes control.[22]

AI is revolutionizing several areas of information technology and it is applied in biomedicine. The applications of AI will grow ten-fold in next 5 years. IBM's Watson Health AI Venture has partnered with 16 cancer institutes to provide computer-aided diagnosis.[23] Companies such as AiCure are taking help of AI applications on smartphones to confirm patients that are taking medications. Berg and InSilico Medicine have started identifying novel drug targets and developing new therapies by leveraging AI.[23] InSilico Medicine enables users to predict their age and gender from the results obtained from blood test, an algorithm is used that can analyze common blood markers like glucose and cholesterol to guess aging. There are other applications, such as Microsoft's How-Old.net and RYNKL, which calculates "youthfulness" with the help of AI by analyzing photographs.[23] Alex Zhavoronkov, the CEO of InSilico Medicine supports the application of AI in disease prevention and management.[23]

According to the experts, deaths occurring from cancer disease can be prevented by at least 22,000 annually using AI diagnostics. Deaths due to diabetes, dementia, and heart disease can also be prevented with AI Applications in health-care diagnostics. The advancement of keen technologies has a capacity to examine the colossal amount of data rapidly with a higher level of exactness; it explores another field of medical research. According to Sir Harpal Kumar, it is a pioneering idea to integrate AI in cancer diagnosis but he cautioned that the infrastructure must be properly designed and it should be in place within the health-care system to make it a reality.[24] Simon Gillespie, the CEO at British Heart

Foundation, acknowledges that AI could be a new weapon to analyze MRI scans and can catch heart disease very quickly and lead to individualized treatments that could save many lives.[24]

According to the researchers at the Madras Diabetes Research Foundation in Chennai India, a smartphone-based device called Remido "Fundus on phone," with the help of AI software, is used for detecting diabetes retinopathy. In fact, the result shows that AI technology had 95% sensitivity and 80% specificity in detecting any diabetic retinopathy.[25] In contrast, the majority of decisions by ophthalmologists had a sensitivity of 0.835 and specificity of 0.981. Google AI researchers are also improving their AI disease detecting software by being trained using a small subset of adjudicated images obtained from ophthalmologists who are specialized in retinal diseases.[25]

A researcher's team led by the University of Cambridge and King Abdullah University of Science and Technology (KUST) has developed a low-cost, semiconducting plastic gadget that can be utilized in medicinal services analysis or in checking a wide scope of health conditions, for example, in surgical difficulties or in neurodegenerative infections.[26] This device allows quickly monitoring of health conditions by measuring the amount of glucose or lactate that is present in saliva, tears, and sweat.[26]

Topological analysis is a new method to diagnose coronary heart diseases. It assists clinician to make more accurate diagnosis. The topological analysis provides a multifeature analysis of heart wall geometry, many of which have not been measured before in cardiovascular imaging. It uses ML that links patient outcome data and predicts whether the patient has significant disease. It delivers an accurate and highly reproducible assessment of the presence of coronary artery disease.[27]

In detecting skin cancer, deep learning convolutional neural network (CNN), which is a form of AI, provides better results than experienced dermatologists.[28] Researchers in Germany, USA, and France had given in excess of 100,000 pictures of malignant melanomas (lethal skin cancer) and benign moles as a training dataset to CNN to detect skin cancer. The CNN has an ability to learn from the images it "sees" and teach itself from what it has learned to improve its performance.[28]

A team of researchers at the University Waterloo, Canada, has developed a smart shirt incorporating sensors to detect heart rate, breathing, and acceleration. The data retrieved from these sensors are then analyzed by AI that assesses changes in aerobic responses and it can predict whether a person is experiencing respiratory or cardiovascular diseases. Thomats Beltrame, who led this research at Waterloo University says, in near future, it is possible to continuously monitor a person's health, even before he realizes that he is need of medical help, and it ultimately impact the quality of life and well-being.[29] Caree Technologies have already developed the smart shirts, called Hexoskin[30], which is currently being used in the research. The research team plans to test these smart shirts on mixed genders and mixed ages, and people suffering from health issues to observe and record how people might react to wear the sensors to check whether their health is failing.[29]

AI and ML have greatly improved in recent years.[31] A new study by Professor Katrilina Aalto-Seetla and Martti Juhola at the University of Tampere in Finland demonstrated that it is conceivable to precisely sort sick cardiovascular cell culture from healthy ones with the utilization of ML and AI, and this technology can also differentiate between genetic cardiac diseases.[31] AI and stem cell technology are combined to learn thrashing cardiomyocytes in the cell culture. The thrashing of a single cardiomyocyte is recorded and then this beating record is given as an input training dataset to ML software and it is taught what disease this cell represented. The ML programming begins figuring out how to isolate various gatherings of cells and can distinguish specific highlights in the beating conduct of every cell.[31]

The researchers have structured a framework to determine some infections to have a high level of accuracy utilizing metabolic markers, which are available in patient's blood at the University of Campinas (UNICAMP) in Brazil.[32] This strategy coordinates a mass spectrometry with a ML algorithm that is equipped for finding a few sorts of examples related to sicknesses of bacterial, viral, fungal, and even hereditary birthplace.[32] This platform utilized infection by Zika Virus as a training dataset and fed it into a computer AI Program. The AI algorithm at the end shows the result whether the patient is infected by Zika virus.[32]

Scientists of GERO Biotech Company have designed Gero Lifespan, which is an iPhone application, which uses smartphone's built in accelerometer to estimate user's life span.[33] Moscow Institute of physics and Technology (MIPT) and researchers from GERO in collaboration have demonstrated that integrating wearable sensors with AI technology can be utilized to create advanced biomarkers of frailty and aging.

This system explores AI potential for health risk assessment based on human physical activity data that are acquired from wearable devices. The

AI algorithms are provided by the physical records and clinical data as an input training dataset from 2003 to 2006 US National Health and Nutrition Examination survey (NHANES). The biologically related motion designs can be identified using AI algorithms and establish their connection to general health and noted life expectancy and ultimately discover the biological age and mortality risks. The combination of ML tools and aging theory can produce better health risk models; it can relieve life span hazards in insurance, and can be benefited in pension planning, and can be used in developing antiaging therapies in future.[33]

Professor Adam Marsh at University of Delaware molecular has started a biotech startup called Genome Profiling LLC (GenPro), which uses novel software incorporated with sophisticated ML techniques and algorithms, takes measurement of methylation patterns in DNA (a cell's genetic code) using next-generation sequencing (NSG) data, and detects DNA methylation patterns that help in uncovering more details about the cellular processes that accelerate spastic cerebral palsy (CP) that could result in improved treatment for CP.[34] The symptoms of CP are jerky movements, muscle tightness, and muscle stiffness that affects 1 in every 323 children making it most common physical disability among children.[34]

The disease "childhood blindness" scientifically known as retinopathy of prematurity (ROP) caused due to abnormal growth of blood vessel near retina; in premature babies, the light-sensitive portion in the back of an eye is a common condition.

Michael Chiang and Jayashree Kalpathy have developed an algorithm that uses AI that utilizes the information of ophthalmologists who are skilled at recognizing ROP and place it into a mathematical model so doctors who probably would not have the real experience can even now help infants to get a convenient and exact diagnosis. The first step in training this AI algorithm was: it was provided with 5000 pictures and was trained to identify retinal vessels. The next step was the algorithm that got trained in identifying healthy and disease retinal vessels. After this, the algorithm's accuracy was compared with the experts and it is found that the algorithm performed better than the most of the expert physicians.[35]

Google researchers have created a device with which the prediction of a person's heart attack can be made. This device looks at scans of the retina and can then able to predict features such as person's blood pressure, his or her age and also if they smoked. These features could be used to predict whether the person has a raised risk of heart disease or a raised risk of cardiac event such as heart attack.[36]

3.6 BENEFITS AND CHALLENGES USING AI

3.6.1 BENEFITS

The application of AI can be observed through a number of initiatives across different sectors like industries, organizations working in health sector, and through government investment.[37] The successful use of AI in health care explored its potential benefits, including optimization of available data, a significant decrease in human errors in analyzing large amount of complex data. AI can possibly make health care progressively productive, moderate, and patient-accommodating. AI helps patients to cope with chronic illness, and also helps in avoiding human bias and errors. AI ultimately speeds up and reduces errors in diagnosis and uses as an assistance tool by many physicians.

3.6.2 CHALLENGES

AI in health care has some important ethical issues that need to addressed, those are who will be held responsible when AI is used in decision-making process. In AI systems, there is a risk of bias present in the data used for training these systems. The use of AI will effect on social isolation of patients and people's sense of dignity in care situations. One of the big challenges is in securing public trust to use AI technology, and the use of AI will affects the roles and skill-requirements of health-care professionals. Other big challenges in using AI technology are lack of accountability from its developers, limitations in patient's data privacy, and the challenges in validation of AI algorithms. Operational and analytical challenges are exists in deployment of AI technology with the real world data. There is a discussion whether AI technology should be used as a toll or as an end product.[37]

3.7 CONCLUSIONS

In today's economy, health care is one of the quickest developing areas. More individuals require reasonable consideration. The utilization of AI in the health-care industry is helping doctors by giving forward-thinking medicinal information from organized and unorganized medical data foundations. AI is outfitted with self-learning and amending capacities, which improve and maximize its performance. AI is helping physicians to maintain good relationship with patients by providing proper diagnosis with accuracy. The researchers in health-care industry are developing smart devices that are helping people with a facility of health risk signals and predicting the health result and ultimately resulting in providing patient-centric care.

KEYWORDS

- **machine learning**
- **artificial intelligence**
- **health care**
- **AI algorithms**
- **artificial neural networks**

REFERENCES

1. The Evolution of Smart Health Care. 2018 Global health care outlook by Deloitte.
2. http://novatiosolutions.com/10-common-applications-artificial-intelligence-healthcare/
3. https://www.forbes.com/sites/forbestechcouncil/2018/01/30/how-ai-is-transforming-the-future-of-healthcare/3/#483d2d661a36
4. https://www.news-medical.net/news/20180522/Health-and-diagnostics-to-soon-be-digitalized-with-advent-of-AI.aspx Ananya Mandal
5. http://www.wired.co.uk/article/future-of-health-technology-ai-wired-health-2018 by Tephen Armstrong. May 22, 2018
6. Louridas, P.; Ebert, C. Machine Learning. IEEE computer Society.
7. Jiang, F.; Jiang, Y.; Zhi, H.; Dong, Y.; Li, H.; Ma, S.; Wang, Y.; Dong, Q.; Shen, H.; Wang, Y. Artificial Intelligence in Healthcare: Past, Present and Future. published on June 21, 2017 by Stroke Vasc Neurol (SVNI)
8. Murff, H.J.; FitzHenry, F.; Matheny, M.E.; et al. Automated Identification of Postoperative Complications within an Electronic Medical Record Using Natural Language Processing. *JAMA* **2011**, *306*, 848–855.
9. Murdoch, T.b.; Detsky, A.S. The Inevitable Application of Big Data to Health Care. *JAMA* **2013**, *309*, 1351–1352.

10. Kolker, E.; Ozdemir, V.; Kolker, E. Hoe Health Care can Refocus on its Super-Customer (Patients, n=1) and Custpmers (Doctors and Nurses) by Leaveraging Lessons from Amazon, Uber, and Watson. *OMICS* **2016**, *20*, 329–333.
11. Dilsizian, S.E.; Siegel, E.L. Artificial Intelligence in Medicine and Cardiac Imaging: Harnessing Big Data and Advanced Computing to Provide Personalised Medical Diagnosis and Treatment. *Curr. Cardiol. Rep.* **2014**, *16*, 441.
12. Patel, V.L.; Shortliffe, E.H.; Stefanelli, M. et al. The Coming of Age of Artificial Intelligence in Medicine. *Artif. Intell. Med.* **2009**, *46*, 5–17.
13. Jha, S.; Topol, E.J.; Adapting to Artificial intelligence: Radiologists and Pathologists as Information Specialists. *JAMA* **2016**, *316*, 2353–2354.
14. Pearson, T. How to Replicate Watson Hardware and Systems Design for Your Own Use in Your Basement. 2011, http://www.ibm.com/developerworks/community/blogs/InsideSystemStorage/entry/ibm_watson_how_to_build_your_own_watson_jr_in_your_basement7? Lang=en (accessed June 29, 2018)
15. Weingart, S.N.; Wilson, R.M.; Gibberd, R.W. et al. Epiemiology of Medical Error. *BMJ* **2000,** *320*, 774–777.
16. Graber, M.L.; Franklin, N.; Gordon, R. Diagnostic Error in Internal Medicine. *Arch. Intern. Med.* **2005,***165*, 1493–1499.
17. Winters, B.; Custer, J.; Galvagno, S.M. et al. Diagnostic Error in the Intensive Care Unit: A Systematic Review of Autopsy Studies. *BMJ Qual Saf* **2012**, *21*, 894–902.
18. Lee, C.S.; Nagy, P.G.; Weaver, S.J. et al. Cognitive and System Factors Contributing to Diagnostic Errors in Radiology. *Am. J. Roentgenol.* **2013**, *2016* 611–617.
19. Neill, D.B. Using Artificial Intelligence to Improve Hospital Inpatient Care. *IEEE Intell. Syst.* **2013**, *28*, 92–95.
20. Bhardwaj, R.; Nambiar A. R.; Dutta D. A Study of Machine Learning in Healthcare. 0730-3157 /17 (c) 2017 IEEE.
21. Taghizadeh, G. Top 5 Companies Revolutionizing Health care with Machine Learning.
22. Freiherr, G. How AI Can Help Patients Manage Diabetes. February 15, 2018.
23. Artificial Intelligence in healthcare predicted to grow tenfold in next five years, Biogerontology Research Foundation, February 24, 2016. www.inslico.com
24. Mandal, A. Incorporation of AI Could Transform Cancer Diagnosis in UK. May 21, 2018.
25. Lovett L. Google Researchers Find Trained AI Detect Diabetic Retinopathy on Par with Experts. March 14, 2018.
26. Researchers Develop Low-Cost Plastic Sensors to Monitor Wide Range of Health Conditions. June 22, 2018. http://www.cam.ac.uk/
27. Anderton, K. Diagnosing Heart disease using AI, An Interview with Dr. Ross Upton. June 22, 2018.
28. Artificial Intelligence Algorithms Appear to be Better at Detecting Skin Cancer. 29 May 2018. http://www.esmo.org
29. Combining Data from Wearable Technology and AI May Help Predict Onset of Diseases. May 16, 2018. https://uwaterloo.ca/news/news/researchers-combine-wearable-technology-and-ai-predict-onset30. www.hexoskin.com

30. Combining Stem Cell Technology and Artificial Intelligence to Diagnose Genetic Cardiac Diseases. 20 June 2018. http://www2.uta.fi/en/news/story/diagnostics-genetic-cardiac-diseases-using-stem-cell-derived-cardiomyocytes, www.nature.com/articles/s41598-018-27695-5
31. Scientists Develop Artificial Intelligence Platform to Diagnose Many Diseases. June 13, 2018. http://agencia.fapesp.br/platform_uses_artificial_intelligence_to_diagnose_zika_and_other_pathogens/28007
32. Researchers Untap Potential of Wearable Sensors and AI Technologies to Predict Bioogical age. March 30, 2018. http://gero.com
33. Researchers Diagnose Cerebral Palsy Using AI and Next Generation Sequencing. June 22, 2018. www.eurekalert.org/pub_release/2018-06/uod-ddd062118.php
34. New AI Algorithm can Accurately Diagnose Cause of Childhood Blindness. May 3, 2018. https://news.ohsu.edu/2018/05/03/ai-better-than-most-human-experts-at-detecting-cause-of-preemie-blindness
35. Mandal, A. Google AI Device Could Predict a Person's Risk of a Heart Attack. February 19, 2018. https:// www.nature.com/articles/s41551-018-0195-0
36. Nuffield Council on Bioethics Outlines ethical issues arising from use of AI in healthcare. May 15, 2018. http://nuffieldbioethics.org/news/2018/big-ethical-questions-artificial-intelligence-ai-healthcare

CHAPTER 4

Healthcare, IoT, and Big Data Support

VIKASH YADAV* and DHANANJAYA VERMA

Department of Computer Science & Engineering,
ABES Engineering College, Ghaziabad, India

**Corresponding author. E-mail: vikas.yadav.cs@gmail.com*

4.1 INTRODUCTION

Healthcare is one of the major sectors in any country, both in terms of the revenue that it generates and the services that it is providing to mankind and also it provides employment opportunities to millions of people. Healthcare sector comprises of technology-enabled health services, multi-facility hospitals, on board healthcare services and telemedicine, multifunction medical devices, and specialized machines.

In the recent years, exponential growth and large amount of investments has been noticed in this sector, taking in consideration the huge amount of revenue that it generates and various types of medical services that it provides.

So, it becomes the need of the hour to power this sector with advanced blooming technologies, for our reference we are focusing on Big Data analytics (BDA) and Internet of things (IoT) in healthcare.

4.2 BIG DATA IN HEALTHCARE

BDA has brought together two different branches of computer science that are Big Data (BD) and analytics, and is collectively designed to deliver a data management approach.

BD is mostly described by the industry professionals as extremely large amount of unstructured and structured data an organization generates. BD has feature of vast size that exceeds the handling capability of traditional information management technologies.

BD has driven the need for development of state of art technological infrastructure and sophisticated tools that are capable of holding, storing, and analyzing huge amounts of structured and unstructured data such as data from biometric applications, medical and text reports, and medical and research images, etc. These data are being generated exponentially from data profound technologies like surfing through the Internet for various activities such as accessing information, visiting, and posting on social networking sites, mobile computing, cloud uploads, electronic commerce, etc.

Now, the corporate industry and various government agencies have realized that there are various unexplored opportunities that can be explored through BDA to improve their work efficiency and reduce unnecessary costs.

Analytics in reference to BD is the process of investigating and analyzing huge amounts of data, coming from various sources of data in all different data formats, in order to provide useful insights that can power decisions at present or in near future. Various analytical technologies and tools such as data mining, deep learning algorithms, natural language processing (NLP) tools, artificial intelligence, and predictive analytics, etc. are used to analyze and contextualize the data. Also various analytical approaches are used to identify inherent patterns in the data, correlations, and anomalies may also get discovered in the integration process of huge amounts of data from different data sets.

BDA helps to draw important conclusions from the data to provide solutions to many important and newly discovered problems. Within the health sector, it provides stakeholders with a blooming new technology and advanced tools that have tremendous potential to move healthcare to a much better level and to make healthcare more economically feasible.

Sometimes there are use cases and tested data for BDA in healthcare that have the potential to help medical researchers to discover causes and treatments for certain diseases. So, that doctors may actively monitor patients for traits of the certain diseases and adequate precautions should be taken before any malefic event to occur and the patient may be provided with personalized care and resources required for the treatment may be gathered before it's too late.

Until the emergence of BDA solutions, researchers and data scientists and government statisticians actively face problems while analyzing huge-sized structured and unstructured data. BDA has minimized the complexity and increased the productivity of the data analysis to greater extent but sometimes problems also occur while using BDA. BDA being a new technology in the market requires state of art technological platform and sophisticated tools that require capital and other functional investments. Technocrats trained with specialized skills to establish are required to operate them.

BD in healthcare apart from being voluminous can also be described through three main characteristics that are as follows:

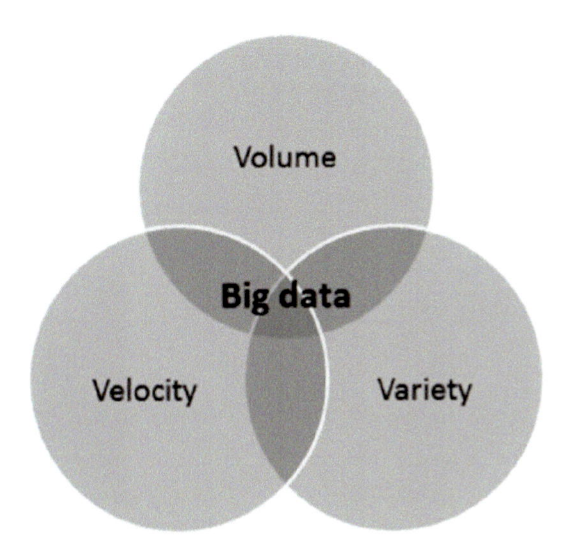

FIGURE 4.1 3V's of Big Data.[1]

Volume can be defined as the amount of data and information generated and consumed by various organizations and individual users. For example, medical imaging (e.g., magnetic resonance imaging (MRI), X-rays, ultrasounds, computed tomography (CT) scans) generates large variety of data with highly level of complexity and much broader scope for further analysis.

Velocity can be defined as the incidence and velocity at which data is being created, hold, stored, and shared among various devices like new technologies available in healthcare are producing billions of data every single minute for a day long.

Variety can be defined as the prevalence of new data types including those coming from machine and mobile sources. Like some healthcare technologies produce "omics" data which are scientifically produced by various sequencing technologies at almost every level of cellular components, from genomics, proteomics, and metabolomics to protein interaction and phenomics.

4.2.1 *INTERNET OF THINGS IN HEALTHCARE*

IoT is an arrangement of multiple physical, electronic, and various sensor-based devices connected to the each other which enables them to store and exchange data among each other.

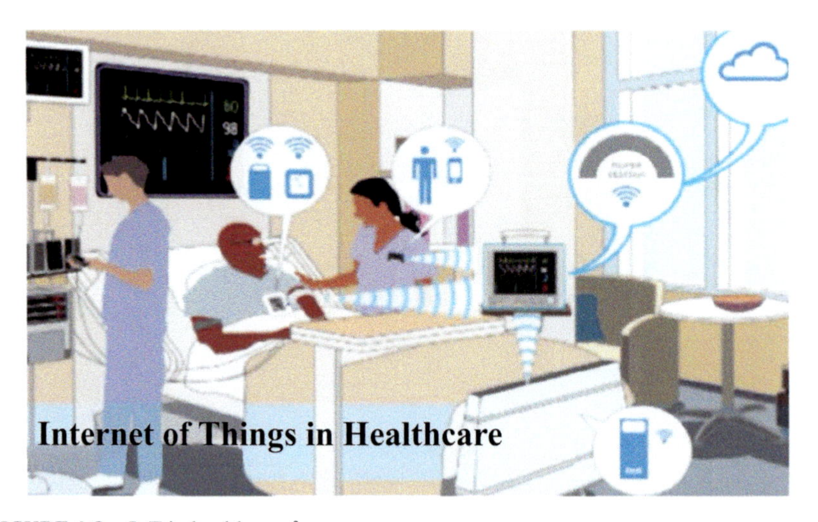

FIGURE 4.2 IoT in healthcare.[2]

With the help of IoT engineers and various developers around the world are able to design and develop devices that are technology-based and can deliver experience-based results in real time. IoT powers the healthcare sector with applications that can be remotely monitored and tele-monitored as well.

According to a research, it is expected that nearly 200 billion devices will be connected to each other by 2020 and the market for IoT-enabled devices may reach up to 6.2 trillion USD and may be more with nearly 38% shares coming from IoT-enabled healthcare devices.

For medical devices and applications to connect with each other to collect exchange data among each other through various IT solutions developed for healthcare, Internet of healthcare things (IoHT) has been developed. It comprises consumerization of various wearable devices which enables personalized monitoring of one's health and many more professional medical devices.

4.3 NEED FOR NEW TECHNOLOGIES

Research and developments in the healthcare sector lead to following factors which in turn accelerated the need for the implementation of new technologies like BDA and IoT as follows:

1. **Higher cost of medical and healthcare services:** BD may reduce the cost of fraud, unnecessary tests, and abuse in the healthcare sector.
2. **Higher demand for overall health coverage:** BD can power various predictive models for better diagnosis and treatment of various diseases quite efficiently by using the data that are being aggregated many times for better results ranging from DNA, cells, to protein-related data.
3. **Increased consumer consciousness regarding one's health:** Consumers are more conscious about their health and demand cost effective personalized treatment and value-based care.
4. **Growing popularity of wearable devices in healthcare:** The modern development in the wearable devices have gained much popularity among youngsters and demand much accurate results.
5. **Growing popularity of handy health monitors for various healthcare issues:** Various development in the monitoring devices have enabled aged patients and handicapped patients to regularly track and monitor their health in real time.
6. **Strategic partnerships among various organizations in the healthcare market for the implementation of various new business models:** The amount of revenue that can be generated

from the healthcare sector make investors to implement new business models to gain more profit and provide better services.

7. **Maintaining Electronic Health Records (EHRs):** For the diagnosis of many diseases, the trend of the patients must be maintained and kept for further research purposes so EHRs are created which comprise of patient-related demographics, medical history, and test-related history and the entry for any special diagnosis or rare diseases which is greatly powered by BD.

4.4 IoT AND BIG DATA SUPPORT

4.4.1 IoT SUPPORT

The IoT provides support to the healthcare sector in the following ways:

1. **Research aid:** IoT powers the medical research with real-world information through real-time data coming from IoT devices and can be analyzed and tested to derive much accurate results

FIGURE 4.3 [3].

2. **Moodables:** These devices are intended to enhance or change the mood of a person. These devices are basically designed and developed through research in neurosciences powered with IoT.

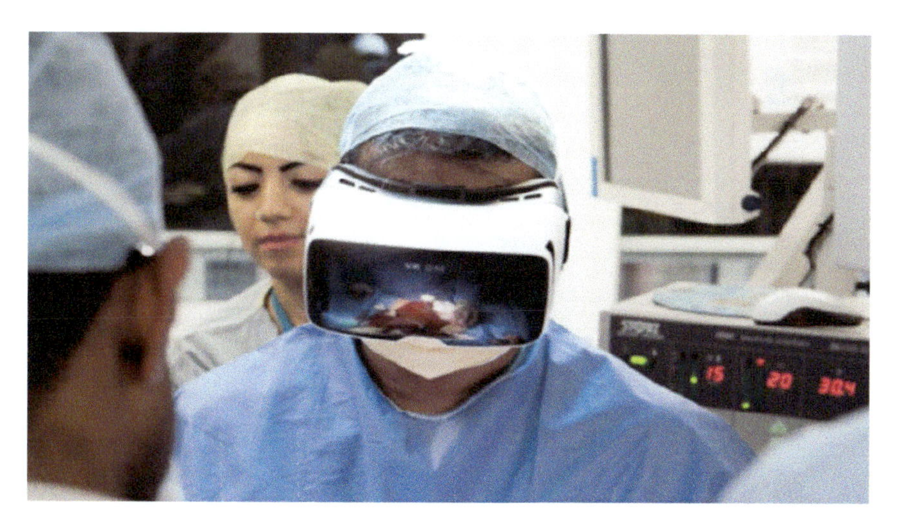

FIGURE 4.4 [4].

3. **Charting:** In healthcare, charting is an important aspect of a patient's disease diagnosis and treatment. This can be automated with the help of IoT by voice commands by the doctor to capture and store data and avoid redundancy. For example, Audemix.
4. **Medical information distribution:** This development of IoT-based applications provide accurate and precise information to the patients and also encourage professional medical practices.
5. **Emergency care:** Through IoT-based applications and devices, the medical aid providers are alerted before any mishappening occurs and also to take appropriate precautions to boost healthcare services.
6. **Reporting and monitoring:** IoT-based devices have made it possible to monitor health in real time and can save lives from various medical conditions like cardiac arrest, varying insulin levels, asthma attacks, etc. and the data can also be shared with the physician in real time.

EHR/Electronic medical records (EMR)/Electronic patient records (EPR)	Puttkammer et al. 2016	HIV
	Landis-Lewis et al. 2015	HIV
	Bruland et al. 2014	Pruritic dermatoses
	Haskew et al. 2015	Maternal and child health
	van Engen-Verheul et al. 2016	Cardiac rehabilitation
	Taggart et al. 2015	Cardiovascular disease an diabetes
	Hoffer et al. 2012	Kidney cancer
	Rahimi et al. 2014	Type 2 Diabetes Mellitus
	Heidebrecht et al. 2014	Public health (immunizatior
	Köpcke et al. 2013	Clinical trial
	Tu et al. 2015	Not mentioned
	García-de-León-Chocano et al. 2015	Maternal and child health
	Weiskopf et al. 2013	Not mentioned
Registry	Adolfsson and Rosenblad 2011	Diabetic
	Rousseau et al. 2014	Population health(vaccinati
Reporting systems	Hirdes et al. 2013	Not mentioned
Hospital information systems	Cohen et al. 2016	Not mentioned
	Herzberg et al. 2011	Medical history forms and stress injection protocols
	Breil et al. 2011	Oncology (urology and haematology)
	Cruz-Correia et al. 2013	Audit trail
Distributed e-healthcare information environment	Wu et al. 2012	Breast cancer
RFID systems	van der Togt et al. 2011	Blood products
per-based records	Adeleke et al. 2012	Inpatient health

FIGURE 4.5 [5].

Emergency Medical Services (EMS)

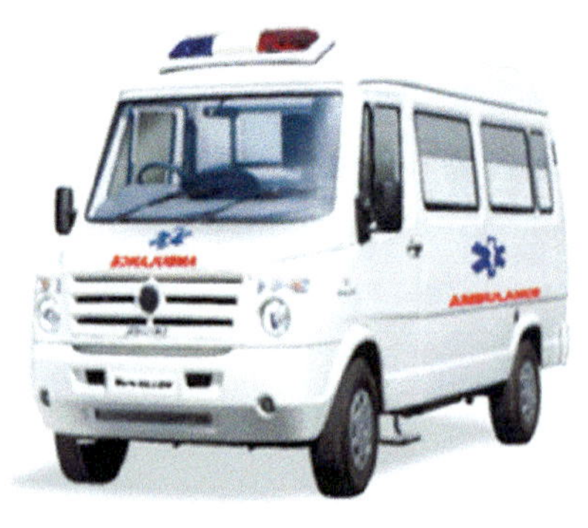
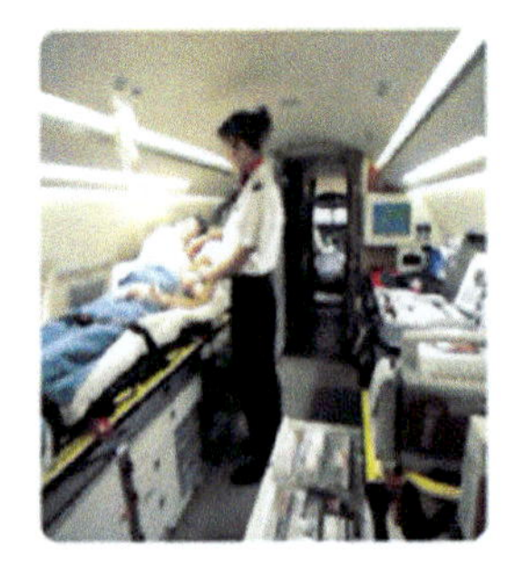

FIGURE 4.6 [6].

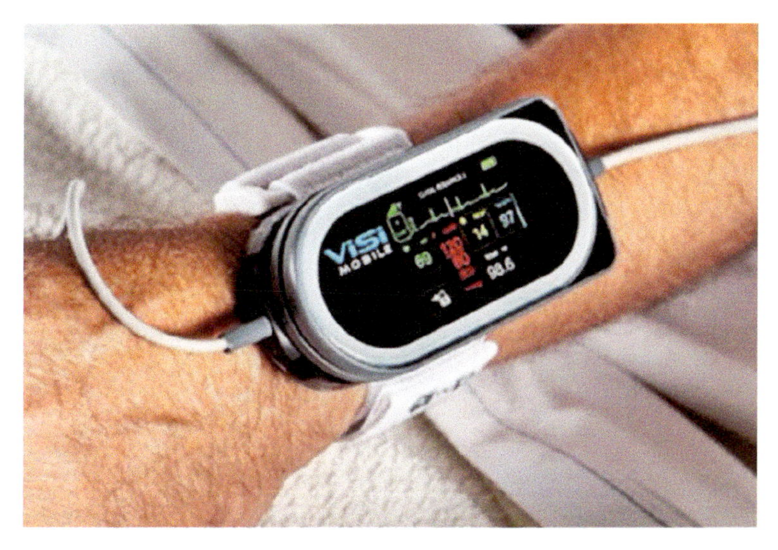

FIGURE 4.7 [7].

7. **Tracking and alerts:** Regular tracking can be possible manually or automated as the IoT-based devices collect and share data in real time so that alerts can be provided according to the varying health of a person.

FIGURE 4.8 [8].

8. **Ingestible sensors:** These are the most unique innovation powered by IoT. They can help monitor and detect any irregularities in body as they are pill sized and can remain inside the body without any harm and benefit healthcare.

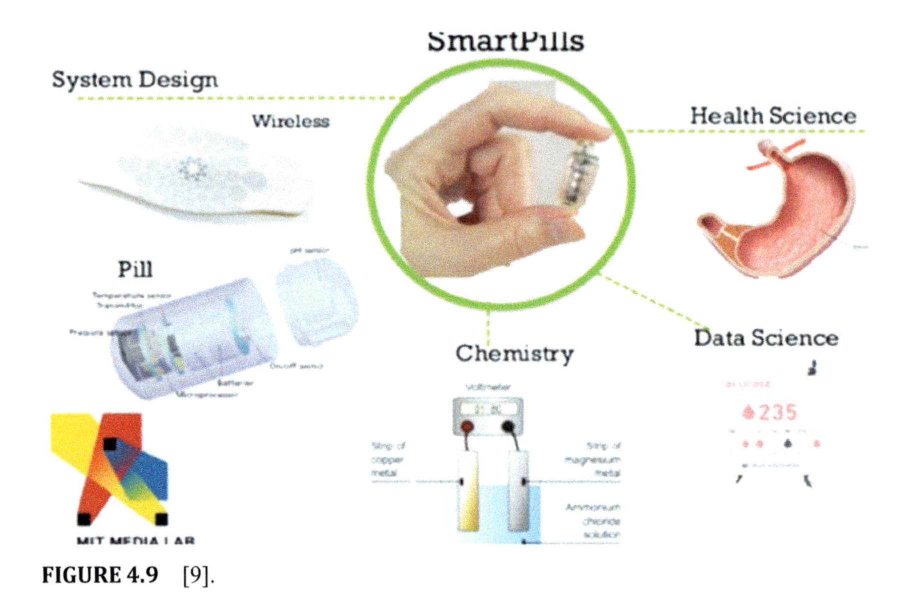

FIGURE 4.9 [9].

9. **Hearables:** These are modern devices that are developed to provide hearing aid in an entirely innovative ways and are powered with technologies like Low Energy Bluetooth, etc.

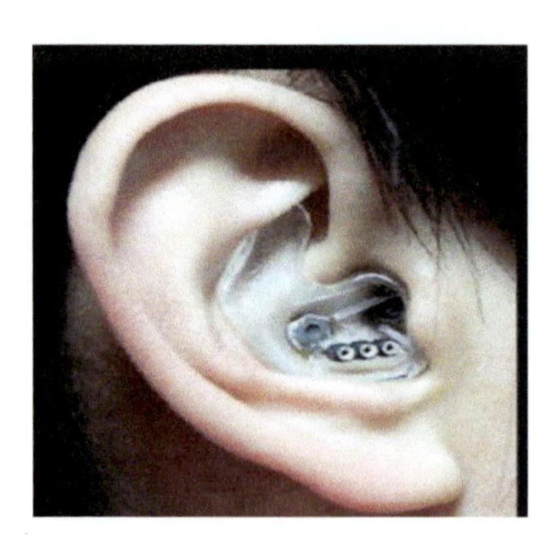

FIGURE 4.10 [10].

10. **Care:** Medical professionals may make better use of their practice in solving better problems with better data and various medical records to study trends of various patients in context of various diseases.

FIGURE 4.11 [11].

4.5 BIG DATA SUPPORT

4.5.1 BIG DATA IN HEALTHCARE: AN ARCHITECTURAL FRAMEWORK

The conceptual framework can use any of the business intelligence tools on any regular system that can be used for a project on healthcare analytics, the reason being that BD processes data across various nodes in small pieces of data that are broken down from large data sets. These data sets are analyzed to explore important insights from them that have the potential to influence the healthcare sector with better decision-making capabilities and many more cost effective diagnosis techniques.

BDA employs various open source platforms like Hadoop, MapReduce, etc. These platforms are programming oriented, complex, and require the deployment of complex infrastructure and technocrats with specialized skills are required to use these tools. The short coming with these technologies is that they are not user friendly and are different from various service-based proprietary tools as they are provided with additional support which is not present in BD tools and platforms.

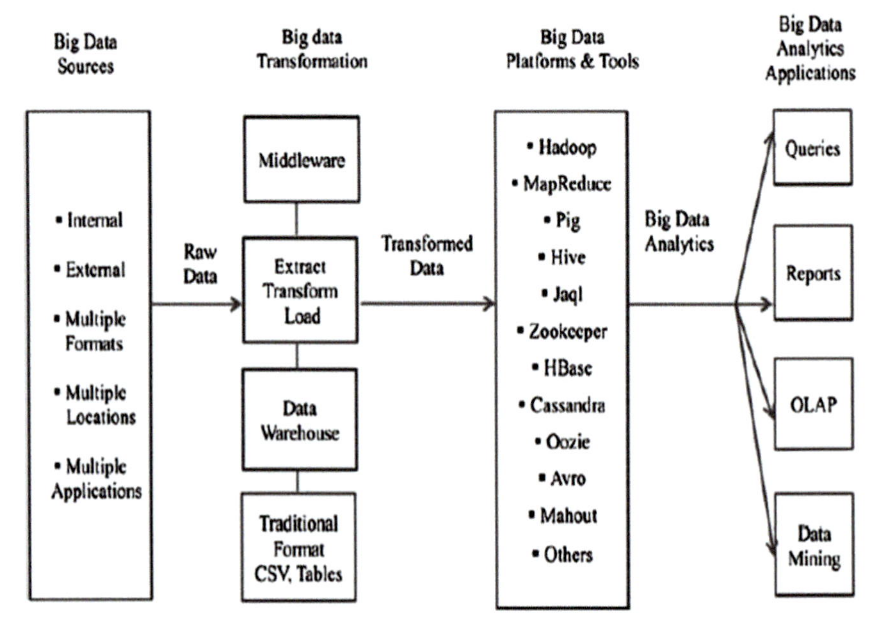

FIGURE 4.12 Conceptual infrastructure of BDA.[12]

The data in healthcare sector is captured, stored, generated, and shared has high complexity within itself as it comes from varied sources like EHRs, public laboratories, various pharmaceutical firms, etc. and from multiple geographic locations from all over the globe, the complexity is high from the very beginning as the data are mostly in unstructured form or may be present in semi-structured or structured form and also the data may be in all different file formats.

The data in healthcare influences the decision-making depending upon the tools and the approach in which the data is fetched. There are many technologies and techniques to analyze, aggregate, and visualize BD in healthcare which are derived out from none other than various mathematical, statistical, and economical techniques. Hadoop (Apache platform) was initially developed for routine functions like aggregating web search indexes belonging to NoSQL techniques. It partitions data sets and then allocates them to various server nodes for further processing and then integrates them to derive final output making them one of the most important among various open source tools and platforms.

Basically it performs functions of both a data analyzer and organizer and possesses ability to process data with varied data structures but possesses high complexity so persons skilled in this are required which are relatively less in number as compared to technocrats in other technologies. Various tools that support the Hadoop distributed platform are listed in Figure 4.13.

Platform/Tool	Description
The Hadoop Distributed File System (HDFS)	HDFS enables the underlying storage for the Hadoop cluster. It divides the data into smaller parts and distributes it across the various servers/nodes.
MapReduce	MapReduce provides the interface for the distribution of sub-tasks and the gathering of outputs. When tasks are executed, MapReduce tracks the processing of each server/node.
PIG and PIG Latin (Pig and PigLatin)	Pig programming language is configured to assimilate all types of data (structured/unstructured, etc.). It is comprised of two key modules: the language itself, called PigLatin, and the runtime version in which the PigLatin code is executed.
Hive	Hive is a runtime Hadoop support architecture that leverages Structure Query Language (SQL) with the Hadoop platform. It permits SQL programmers to develop Hive Query Language (HQL) statements akin to typical SQL statements.
Jaql	Jaql is a functional, declarative query language designed to process large data sets. To facilitate parallel processing, Jaql converts "'high-level' queries into 'low-level' queries" consisting of MapReduce tasks.
Zookeeper	Zookeeper allows a centralized infrastructure with various services, providing synchronization across a cluster of servers. Big data analytics applications utilize these services to coordinate parallel processing across big clusters.

FIGURE 4.13 Platforms and tools for BDA in healthcare.[12]

4.5.2 METHODOLOGY

The stages of methodology are discussed in Figure 4.14. Firstly, a concept statement is developed by the team for BDA in healthcare team and after that the importance of project is described and is further sent for its approval. If approved it proceeds for the proposal development stage. This stage addresses the need and benefits the project delivering to the stake holders and the approaches and how they are used for BDA as it is much complex than traditional data

Warehousing and BI tools have costlier implementation than others. Also project background and prior research related to the projects are discussed.

Step 1 Concept statement

Step 2 Proposal
- What is the problem being addressed?
- Why is it important and interesting?
- Why big data analytics approach?
- Background material

Step 3 Methodology
- Propositions
- Variable selection
- Data collection
- ETL and data transformation
- Platform/tool selection
- Conceptual model
- Analytic techniques

-Association, clustering, classification, etc.
- Results & insight

Step 4 Deployment
- Evaluation & validation
- Testing

FIGURE 4.14 Methodology.[12]

Third stage is the implementation stage of the methodologies figured out from the first and second stages, and the concept statement is broken down to small and easily processable propositions. In this stagethe identification of independent and dependent variables if the data sources are done. The data are transformed and then the procedure proceeds to tool/

platform selection and evaluation selection is done. Then after the selection of appropriate tool BDA techniques are applied to these large data sets and then various important conclusions and vital insights are drawn out as a result of the procedure. These insights greatly empower decision-making in the healthcare sector, and lastly, the outcomes and results are tested and validated are finally present in the last step. The models and their results have been tested, validated, and handed over to the stakeholders in the healthcare sector; in this staged procedure, various feedback sessions are also arranged at the end of every stage to minimize the chances of errors.

BD provides support to the healthcare sector through the following ways:

1. **Bioinformatics applications:** Research in bioinformatics analyzes various variations in the biological system up to molecular level. It is necessary to store and analyze data in a time efficient manner to align with the current trends of healthcare. With the help of modern sequencing technologies, the genomic data can be acquired in very less time. BDA help bioinformatics applications to create data repositories, and also it provides infrastructure for computing, and other data manipulation tools to efficiently analyze biological data and draw valuable insights from it.
2. **Improved security and reduce fraud:** Some research results also indicate that the data in the healthcare sector is 200% more likely to be breached than any other sector and any harm to the personal and medical data would result in catastrophic consequences because it possesses great market value for the cyber criminals in the black markets. By the use of BDA, many threats can be prevented by identifying and analyzing any change in the network traffic and any other type of attack also it prevents many inaccurate insurance claims in a systematic way. It empowers various encryption techniques, network firewalls, processing of insurance claims, enabling patients to get better returns on their claims, and caregivers are paid faster. For example, according to the Centre for Medicare and Medicaid Services Big Data analytics helped them to save more than 210.7 million USD in just a year by preventing frauds and boosting their security standards.

   BD when collectively used with the IoHT can help a person to track various user statistics and other important information. IoT-powered wearable devices are able to detect sleep duration of a person, heartbeat in real time, calories burned, etc. and also there are devices which monitor a patient's blood pressure, blood glucose levels. This real-time tracking of body will allow patients to take appropriate precautions regarding their health when necessary.

3. **Predictive analysis:** BD helps to analyze use cases and tested data for BDA in healthcare that have the potential to assist researchers to discover causes and treatments for certain diseases. Such that doctors may actively monitor patients for traits of the certain diseases and adequate precautions should be taken before any malefic event to occur and the patient may be provided with personalized care and resources required for the treatment may be gathered before it's too late.
4. **Human errors prevention:** Sometimes healthcare professionals may unintentionally prescribe a wrong medicine or wrong doses for a drug. These types of errors can be minimized by the use of BD as it can be used to analyze data and the corresponding prescribed treatment. By this, the loss due to human errors can be minimized and many lives can be saved. Hence, by the use of BD, many busy healthcare professionals may work more efficiently and reliability toward healthcare is boosted.
5. **Building diagnostic machines:** These are the innovation in which the medical machines diagnose the disease and intelligently interpret the results by itself so that the doctors are alerted for any abnormality if found in the results. For example, in case of X-rays, MRI's, and ultrasound images the machine with the aid of BD can analyze results as well as report for any abnormalities.
6. BDA performs some unique functions such as it provides solutions to questions of type "how" and "why" despite of traditional data warehousing tools which provide solutions to only specific problems of type "what" and "where".
7. **Sources of Big Data in healthcare:** In the healthcare sector and among various IoT-enabled devices the sources of BD include:

   - **Internet transaction:** Billions of online purchases, stock trades, social networking exchanges, Internet searches, etc. are being performed on internet in real time, including infinite digital transactions. Each result in the creation of multiple data points collected from online retailers, bank firms and banking services, social networking, and search engines, etc.

- **Mobile devices:** Among billions of mobile devices present around the world, each one of which is generating data through call, text, and various multimedia messages. Mobile devices also transmit location data. Each social media update, tweet, blogs, and user comments create multiple data points that are mostly unstructured and can also be semi-structured which are called as data exhaust.
- **IoT-based devices and other electronic devices:** Electronic devices of various types that include servers and other related hardware devices, modern energy meters and other sensor-based devices, monitors, etc. they all mostly generate data in unstructured form and also in semi-structured form.
- **Online streamed data:** Various IoT-based monitoring devices, handhelds, telehealth devices, and other sensor-based smart devices generate data all in different formats and also considerable amounts of real-time data for further analysis by various healthcare systems.
- **Data coming from biometrics:** These types of data include data collected from various biometric devices like finger print scanners, retinal scanners, face recognition systems, iris scanners, palm recognition systems, etc.
- **Data collected through Internet surfing and social networking sites:** It contains data that is generated through search engines, surfing through the Internet, and data generated from many social networking sites.
- **Clinical publications-related data:** Text-based publications that include medical research and other reference material, also clinical practice guidelines and other health products-(e.g., drug information) related data.
- **External data in healthcare:** This type of data includes financial reports, billing data, job scheduling data, administrative data, etc.
- **Clinically generated data:** More than 79% of health data is unstructured in the form of medical documents, medical images, healthcare records, and all coming from new data sources.
- **Value-based healthcare delivery**: The delivery of value-data completely depends upon the coordination between medical treatment and financial success. To draw important conclusions and valuable insights from the correlation between efficiency

and treatment cost, further analytics are needed to be performed on the data present in integrated and heterogeneous form.

- **US legislation:** The US Healthcare Reform, also called as Obamacare supported the functioning of techniques like BDA and HER which as a result significantly influenced the international market for emerging technologies.
- **Incentives:** BD also helps to align incentives of medical systems with their outcomes and also several incentives such as Care Organizations (ACO) (Centre for Medicare and Medicaid Services 2010), or Diagnose-related Groups (DRG) (Ma Ching-To Albert 1994) have been implemented in order to reward quality over quantity of treatments.

8. **Big Data in image processing:** One of the important sources of medical data are medical images generated through various imaging techniques like CT, MRI, X-rays, molecular imaging, ultrasound, photo acoustic imaging, fluoroscopy, positron emission tomography–computed tomography (PET-CT), etc. and require large data storage infrastructures. These images hold a wide spectrum of different image acquisition methodologies. For example, structures of blood vessels can be visualized using MRI, CT, ultrasound, and photo acoustic imaging, etc.

   A type of medical imaging for which the implementation of analytics is difficult is molecular imaging. It is a noninvasive technique of cellular events which can help diagnose diseases like cancer.

   Another type of medical imaging is microwave imaging that has the ability to map electromagnetic wave scattering from the contrast in the dielectric properties of different tissues. It has both functional and physiological information encoded in the dielectric properties which can help differentiate and characterize different tissues and/or pathologies.

   - **Big Data in signal processing:** Healthcare devices and monitors generate data with large volumes and with a great velocity and also the physiological signal show spatiotemporal complexity. Data in situational context along with these physiological signals are needed to be rooted to ensure the effectiveness of checking and projecting systems. The continuous data generated from these systems has been utilizing telemetry and constant physiological time series monitoring to perk up patient care.

9. **Streaming data analytics in healthcare:** The efficient use of constant waveform when further analyzed through applied analytical disciplines (e.g., statistical, quantitative, contextual, cognitive, and predictive) to impel out decisions for patient care is called as streaming data analytics in healthcare.
10. **Genomic data:** Due to the advancement in sequencing technologies that cost to sequence a genome with more than 30K genes have greatly reduced and the efficiency is improved by more than five orders of magnitude and genome-wide analysis has utilized microarrays to analyze various traits so that the treatment of diseases like Crohn's disease and age-related muscular degeneration are made possible. BD applications in genomics majorly focus on pathway analysis in which functional effects of genes is differentially expressed in an experiment or gene set of particular interest are analyzed.
11. **Data aggregation:** Integration of data coming from different sources and the process of developing consistency in data and making the data standardized is defined as data aggregation and medical data being of higher complexity is majorly present in unstructured form needs to be simplified to lower complexities through data aggregation.

4.6 TECHNOLOGY-ENABLED COST OPTIMIZATION

In this section, we will explore how new technologies optimized cost and increased accuracy of results and also how the cost of implementation of new technologies is affecting the healthcare sector across various countries.

In developed countries like the United States the healthcare services are costlier than ever before and as per the report of Mc Kinsey the cost of money invested on healthcare in the United States contributes to nearly 17.6% of GDP which is approximately $600 Billion more than the estimated costs, and this calls for some cost optimizing technologies like BDA which not only reduce the healthcare expenses but also improve the level of services that the healthcare sector is delivering to its stakeholders as the healthcare services are also not up to the mark in terms of quality and efficiency and the ever increasing population demands more of personalized healthcare and healthcare services that save time and money as well and provide them with accurate results.

Hence, there is a need for technology-enabled cost and services optimization.

BDA and IoT-enabled devise optimize cost and services in the health-care sector in the following ways:

1. **Optimizing cost due to fraud waste and abuse in the health-care sector:** One of the main reasons of inflation in the healthcare sector is the cost due to frauds and abuse in the healthcare sector. It is evident that predictive analysis has prevented costs of worth millions, various organizations like Centre for Medicare and Medicaid Services implemented predictive analysis and saved more than $210.7 million in a year itself and the implementation of predictive modeling is used by many organizations to identify illegal claims and are able to save millions of bucks in a year.
2. **Disease diagnosis optimization:** Some of the chronic diseases like rheumatoid arthritis, etc. are greatly powered by BDA as it helps to maintain the trend of the patient's disease track record and also help to analyze the data to gather important insights related to the disease and to provide specialized care to the patient.
3. **Medication with minimized human errors:** Human errors in the medication may lead to catastrophic circumstances and may also lead to even death of the person. By the use of BDA, the diagnosis can be made more précised and accurate which reduce the extent of errors to very less or negligible amount.
4. **Monitoring of patients in real time:** Data coming from various IoT devices in real time can be analyzed in the same time to derive out predictive insights regarding the patient and to provide the patient with tailored medication and personalized care. For example, MedAware, an Israel-based healthcare company has implemented BDA and are able to optimize cost approximating to $21 billion in a year.
5. **Reduction in hospital wait time:** BDA can also be implemented along with IoT devices to gain insights from patient's data to reduce hospital wait time as it can be used by hospitals to predict the number of patients they are expecting in a month or a week and also to predict the patients with high risk situation to reduce the number of readmissions. Some examples are many hospitals in the United Kingdom are using BDA to predict the number of patients they expect in a month.

4.7 CHALLENGES IN INTRODUCTION OF NEW TECHNOLOGIES TO HEALTHCARE SECTOR

1. **Data security and privacy:** There are much chances in IoT that the data security and privacy may get compromised as the data being sent and received among various IoT-enabled devices lacks various standard data transmission protocols and also ambiguity may arise regarding data ownership and its regulation. These factors make data prone to hackers and crackers and result of which the personal health information (PHI) may get compromised and both patient and doctor may have to suffer from the loss due to the same.
2. **Hindrance in integration of multiple devices:** The developers of various IoT-enabled devices do not use standard protocols for communication and thus the devices may face hindrance while transmitting and storing data which as a result slows down the process of data aggregation among various devices and thus reduces the scalability factor of IoT-enabled devices in healthcare.
3. **Data accuracy and overload:** Due to the nonuniformity of protocols for data transmission among various IoT-enabled devices, data accuracy is compromised and some devices also face issues like data overload, fact being that IoT devices may result in accumulation of terabytes of data which can be further analyzed to gain insights.
4. **Costs:** The higher cost of healthcare services in developed countries makes them difficult for common man to access them so it gives rise to the concept of medical tourism in which the person travels to developing countries for treatment and can save up to 90% of its expenditure on healthcare services. But there is a huge scope through IoT- and BD-powered devices to cut off medical expenditures and healthcare costs.
5. **Closing the loop for drug delivery:** The loop for drug delivery can be closed by combining data from IoT-enabled devices and data analytics and developing devices that can automatically respond to the corresponding changes. For example, applications like artificial pancreas. Also, it has the potential to develop automated drug delivery systems and other sensor data-based and data-driven applications. Medical devices or smart phone-based health apps

powered with BDA can automate the systems for drug delivery and give recommendations for various diseases.

6. **Lack of digitalized health data:** A very small amount of medical data and clinical information is available in digital format and is ready to be used for analytics.
7. **Lack of unified health data:** The health data should be unified among various data sets for patients across different hospitals.
8. **Data storage:** Data stored in different and distributed data silos makes it difficult for further analytics to happen and make analytics highly unstable.
9. **Organizational storage of data:** Lack of incentives may lead to problems in cooperation across different organizations and even in interdepartmental data exchange.
10. **High investments:** As the BD healthcare applications require standard quality of data available in large quantities for further analytics, collection of these data sets is not only costlier and also time consuming.
11. **Missing business cases and confusing business models:** Any new technology is needed to be aligned with a successful business case for its successful implementation. So, there should be a clarity needed in defining the stakeholders of the technology in both ways that the payer, consumer, and the driver of the implementation of the new technology for proper implementation.

4.8 CONCLUSION

Healthcare sector is of uttermost importance in any country in terms of both, the revenue that it generates and the services that it is providing to mankind. It comprises of technology-enabled health services, multi-facility hospitals, on board healthcare services, telemedicine, and specialized machines, etc. and also the large-scale employment that this sector provides makes this sector of prime consideration. So, this sector needs to be powered with technology-enabled health services and employ the usage of new technologies like BDA and IoT in healthcare.

In this chapter, we have majorly focused on two technologies that are BDA and IoT in healthcare.

BDA has brought together two different branches of computer science that are BD and analytics, and is collectively designed to deliver a data management approach. BD is mostly described by the industry professionals as extremely large amount of unstructured and structured data an organization may hold, generate, and share.

BD in healthcare can be described through three main characteristics that are as follows:

Volume is the amount of data and information generated and consumed by various organizations and individual users. For example, medical imaging (e.g., MRI, X-rays, ultrasounds, CT scans, etc.).

Velocity is the frequency and speed at which data is being created, hold, stored, and shared among various devices like new technologies available in healthcare are producing billions of data every single minute for a day long.

Variety is the prevalence of new data types including those coming from machine and mobile sources.

BD has feature of vast size that exceeds the handling capability of traditional information management technologies such as previously exiting traditional data warehousing tools and BI tools. It has driven the need for development of state of art technological infrastructure and sophisticated tools that are capable of holding, storing, and analyzing large amounts of structured and unstructured data such as medical, text reports, research images, Internet transaction, mobile devices, IoT-based devices, online streamed data, data coming from biometrics, data collected through social networking sites, clinical publications-related data, etc.

These data are being generated at large scales and in comparatively less time from data profound technologies like surfing through the Internet for various activities such as accessing information, visiting and posting on social networking sites, mobile computing, cloud uploads, electronic commerce, etc.

Analytics in reference to BD is the procedure of investigating and analyzing BD coming from various sources of data in all diverse data formats, in order to provide useful insights that can power decisions at present or in near future.

Several analytical technologies and tools like data mining, deep learning algorithms, NLP, artificial intelligence, and predictive analytics, etc. are used to analyze and contextualize the data. Also, various analytical approaches are used to identify inherent patterns in the data, correlations, and anomalies may also get discovered in the integration process of huge amounts of data from different data sets. BDA being a new technology in the market require state of art technological platform and sophisticated tools that require capital and other functional investments. Technocrats trained with specialized skills to establish are required to operate them.

IoT is an arrangement of multiple physical, electronic, and various sensor-based devices connected to each other which enables them to store and exchange data among each other. For medical devices and applications to connect with each other, IoHT has been developed. With the help of IoT engineers and various developers around the world are able to design and develop healthcare devices that are technology-based and can deliver experience-based results in real time.

BD when collectively used with the IoHT can help a person to track various user statistics and other important information. IoT-powered wearable devices are able to detect sleep duration of a person, heartbeat in real time, calories burned, etc. and also there are devices that can monitor a patient's blood pressure, blood glucose levels. This real-time tracking of body will allow patients to take appropriate precautions regarding their health when necessary. IoHT along with BDA empowers various healthcare facilities such as medical research, Moodables, charting, medical information distribution, emergency care, reporting and monitoring, tracking and alerts, ingestible sensors, hearables, bioinformatics applications, improved security and reduce fraud, predictive analysis, building diagnostic machines, image processing, BD in signal processing, etc.

Some of the challenges that these new technologies are facing in their hassle free implementation are data security and privacy in data coming from IoT devices, hindrance in integration of multiple devices, data accuracy and overload, closing the loop for drug delivery, lack of digitalized health data, lack of unified health data, data storage, high initial investments, missing business cases and confusing business models for implementation, etc. So, when huge amount of data coming from IoT devices and other sources of BD in healthcare are processed and analyzed through BDA they can help to draw useful insights and enhance the decision-making capabilities of this sector and move the healthcare sector to a new level of much higher productivity and personalized and error free healthcare services.

Taking in consideration the higher cost of implementation of these technologies, these require complex infrastructure, persons with specialized skills to install, configure, and work on these technologies but this implementation is endorsed by government agencies as they are realizing its long-term benefits and more accurate and precise results can be implemented in a country in a more efficient way. Also, the research and educational institutions need to educate and train young minds in order to use and explore the unleashing boundaries of these technologies and the never ending support it can provide to mankind also can curb the cost of healthcare services in developed and developing nations as well so that the high cost incurring healthcare services can be afforded by common man.

The implementation of these technologies is also time efficient as it saves lots of time that was previously wasted in treatment and disease diagnosis. Hence, BDA and IoT are a source of never ending support for healthcare sector and other biomedical services.

REFERENCES

1. https://bigdataldn.com/big-data-the-3-vs-explained/
2. https://readwrite.com/2018/01/13/internet-things-healthcare-possibilities-challenges/
3. https://yourstory.com/2018/08/market-research-importance-healthcare/
4. https://www.solutionanalysts.com/blog/5-iot-applications-that-will-change-the-face-of- healthcare/
5. https://www.researchgate.net/figure/Distribution-on-the-forms-of-healthcare-records-in- this-review_tbl2_318897881
6. https://www.indiamart.com/proddetail/emergency-medical-services-4795931133.html
7. https://gajitz.com/touchscreen-monitor-lets-doctors-monitor-patients-remotely/
8. https://www.networkworld.com/article/3295905/internet-of-things/lab-makes-data-sharing-easier-so-medical-iot-devices-can-be-smarter.html
9. http://medgizmo.info/news/ingestible-sensors.-medgizmo-update
10. https://www.openpr.com/news/673708/Hearable-Devices-Market-Report-Forecasts-Strong-Growth-by-2022-Scalar-Market-Research.html
11. https://www.jehangirhospital.com/about-us/hospitals-overview
12. Raghupathi, W.; Raghupathi, V. Big Data Analytics in Healthcare: Promise and Potential. *Health Inf. Sci. Syst.* **2014,** *2*(3). http://www.hissjournal.com/content/2/1/3

CHAPTER 5

GROCD: Novel Fuzzy Rules Based on Efficient Clustering and Classification of BDNF with Type-2 Diabetes Mellitus

DHARMAIAH DEVARAPALLI[1*], PHANIGRAHI SRIKANTH[2], and AHMED A. ELNGAR[3]

[1]*Department of Computer Science and Engineering, Shri Vishnu Engineering College for Woman, Bhimavaram, Andhra Pradesh, India*

[2]*Department of Computer Science and Engineering, GMR Institute of Technology, Rajam, Andhra Pradesh, India*

[3]*Faculty of Computers and Artificial Intelligence, Beni-Suef University, Beni-Suef City, Salah salem str., 62511, Egypt*

[*]*Corresponding author. E-mail: devarapalli.dharma@gmail.com*

ABSTRACT

Today most of the peoples are suffering diabetes mellitus. Working on type-2 diabetes mellitus of brain-derived neurotrophic factor (BDNF) gene data elaborated as data mining and machine learning using developed applications, the main objective of this paper is to propose the methodological framework design of GROCD as "Novel fuzzy rules based on efficient clustering and classification of BDNF with type 2 diabetes mellitus decision support tool based on proposed framework developed as examination of the process based on identification of the patients," which are sustained important problems for identification of diabetes patient. Using this framework, the type-2 diabetes mellitus can be identified in patients.

The most familiar approach is the novel fuzzy rules with an efficient clustering algorithm with proposed similarity measure, information gain, and similarity based on defined clusters. Classification algorithm is *k*-nearest neighbor (kNN) with proposed similarity measure sigmoid function with Euclidean space using classification of patients, either positive or negative. These methods using constructed as framework is GROCD and the framework is based on identification of patient's severity as low or high. This framework is based on detection of cases as positive and negative; case studies and results show that it is most effectively identified in patient management.

5.1 INTRODUCTION

Brain-derived neurotrophic factor (BDNF) gene with type-2 diabetes mellitus diagnosis data applied into different types of clustering and classification algorithm based on positive and negative patients and classify patient severity also as low, medium, and high.

5.1.1 BRAIN-DERIVED NEUROTROPHIC FACTOR

BDNF gene is a neurotrophic factor that affects the brain. In this BDNF gene, mainly there are two types of mechanisms: TrkB and LNGFR; BDNF with SNP based on the investigated mRNA genes that are extracted. BDNF 3D structure is shown in Figure 5.1.

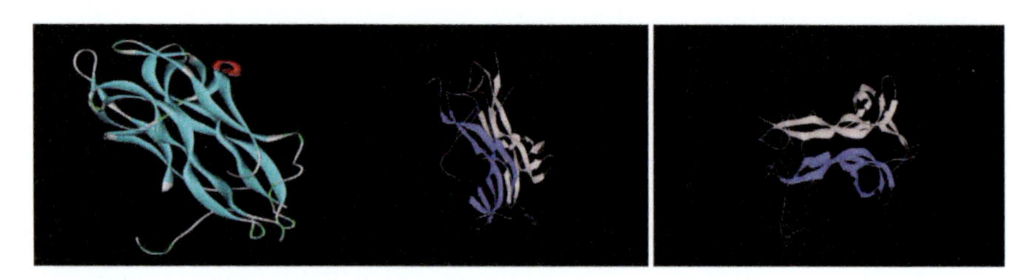

FIGURE 5.1 3D structure with BDNF.

5.1.2 BDNF GENE WITH TYPE-2 DIABETES MELLITUS

A presentation of BDNF gene with type-2 diabetes mellitus with diagnosis data of different attributes and patient records with finding attributes of diagnosis-related data is shown in Figure 5.2.

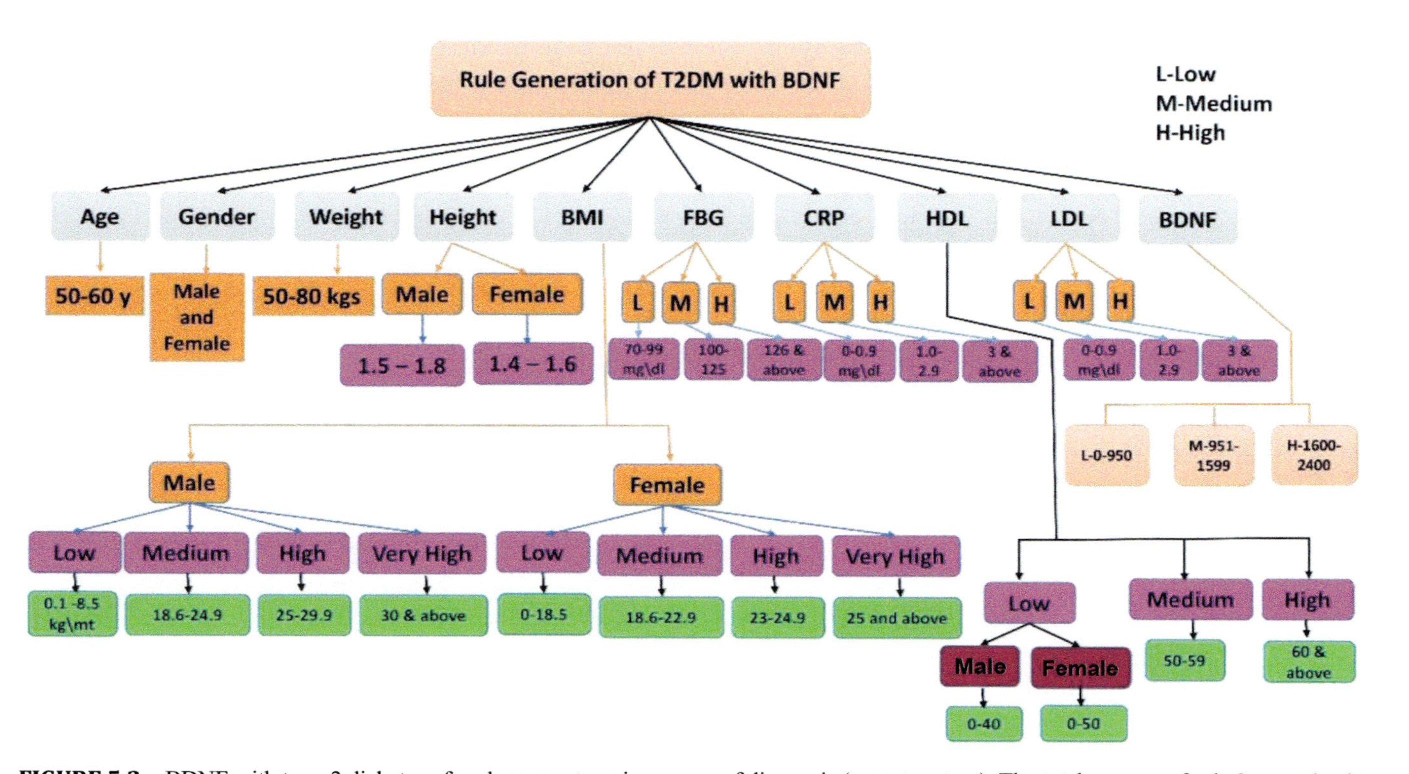

FIGURE 5.2 BDNF with type-2 diabetes of each parameter wise ranges of diagnosis (expert system). The total average of rule 3: completely 30 parameters of the diabetes dataset and 2 classes of the data.

5.1.3 BDNF GENE EXPERT SYSTEM

BDNF gene with expert design system defines each attribute wise ranges of low, medium, and high with 10 attributes and the class is positive and negative with low, medium, high, and very high, based on identified patient severity. A detailed description is shown in Figure 5.2.

5.2 RESEARCH PROBLEM

5.2.1 RESEARCH OBJECTIVE

This main objective of this chapter is to define novel fuzzy rules of BDNF gene with type-2 diabetes of different rules and defining clustering proposing measures based on clusters generated and reduced matrix. After getting clustering based on the build reduction matrix, the classification algorithm kNN is applied based on the prediction of positive and negative results that define patient severity.

5.2.2 RESEARCH CHALLENGES

The important challenges for the collection of data that BDNF with type-2 diabetes mellitus face are presented as follows:

1. Identification of specific attributes pf BDNF with type-2 diabetes and identification of attributes of the gene
2. Expert design system of BDNF with type-2 diabetes
3. Generating number of rules with fuzzy and fuzzy membership function of low, medium, and high conditions generating rules with BDNF with type-2 diabetes
4. BDNF with type-2 diabetes of each, rule wise or query wise, defined clusters or generating clusters.
5. Generating clusters with BDNF with type-2 diabetes based on defined dimensionality reduction matrix of reducing the data is also a very important challenge.
6. Classification of kNN algorithm proposing new similarity measure and validating is also a very important challenge.
7. Defining patient prediction and patient severity (low, medium, and high) are also very important challenges.

5.3 RELATED WORK

Research work followed by literature of medical diagnosis based on the identification of patients and define patient severity based on build framework as predicted patients. Most research work of Korely et al. discuss BDNF with patient diagnosis and patient's behavior.[1] The paper of Lee et al. discusses BDNF with specific parameter based on defined patient severity in detail.[2] The paper of Patil et al. discusses patient severity of type 2 diabetes with hybrid model using predicted diabetic patients.[3] The paper of Dharmaiah et al. discusses BDNF with type-2 diabetes data based on built expert system each attribute wise and specific and accurate ranges.[4,5] The paper of Varma et al. discusses patient identification with principal component analysis with SLIQ decision support system and computational intelligence techniques predict patients.[6,7,19]

5.4 METHODOLOGY

5.4.1 FUZZY RULE: NOVEL FUZZY RULES OF BDNF WITH TYPE-2 DIABETES MELLITUS

Novel fuzzy rules with BDNF with type-2 diabetes mellitus algorithm

Input: Dataset is BDNF with type diabetes patient's data.The numerical value attributes and class of each row and column of identifying each record based on decision taken to predict the diabetes dataset.

Output: Fuzzy rules with BDNF with type-2 diabetic patient accurate decision of identified.

Step 1: Select all diabetic and nondiabetic dataset build in numerical order.

Step 2: After building the dataset, refer to each attribute, record-wise identify the ranges, and condition-wise check step 3.

Step 3: Then each attribute wise is identified as low, medium, and high.

Step 4: Then apply each condition of the computed novel fuzzy rules as if and then.

If condition satisfies both LHS and RHS, compute fuzzy rules of each case rule wise.

Step 5: Repeat step 3 and step 4 of each patient.
Step 6: Stop.

BDNF with type-2 diabetes data with define as rules as different conditions of design as expert system. Novel fuzzy rules are defined as depending knowledge base with depending designing each attribute and class wise at a range of low, medium, high, and very high.[6,7,9]

Fuzzy rules of attribute with condition and then attribute with condition followed by developed and designed rules with expert system in the ranges of low, medium, and high.

1. BDNF with type-2 diabetes mellitus (BDNF-T2DM) with diabetic condition for male.
2. BDNF-T2DM with diabetic condition for female.
3. BDNF-T2DM with prediabetes condition for male.
4. BDNF-T2DM with prediabetes condition for female.
5. BDNF-T2DM with diabetes-free condition for male—healthy condition (no diabetic condition).
6. BDNF-T2DM with diabetes-free condition for female—healthy condition (no diabetic).

Rule 1: Low condition

If (LHS) then (RHS) = attribute versus class = true positive.

If

(Age (0–80 and above) and gender (male and female is 1 and 0) and material status (material is married and unmarried is 1 and 0) and weight (50–80 kg) and height (male is 1.524 to 1.8288 and female is 1.4224 to 1.6460) and BMI (male is 0–18.5 and female is 0–18.5) and HbA1c (4.0–6.0) and cholesterol (>200 mg/dl) and amylase (40–140 so low is 40–78) and creatinine (0.7–1.3) and insulin test (yes or no (1 or 0)) and urine test (yes or no (1 or 0)) and FBG (70–99) and HDL (>1.33) LDL (2.6 to 3.33) and triglycerides (>150) and fasting sugar of 75 g fasting (60–100) and 1 h > 200 and 2 h > 140 and random sugar (79–140) and PP blood sugar (below 140 (<140)) and blood URA (children 5–18, youth 6–20, adult 8–20) and uric acid (yes or no) and HbA1c 4.0–6.0 and pregnant (yes or no) and DPF < 0.4 and BP (90/60) and thyroid (yes or no) and skin flood (ranges is for all 30–35%) and smoking (yes or no) and depression and stress (yes or no) and urine test (yes or no)) *or*

Then class diabetes (P or N)

Rule 1 defined as truly P or true N

This case is applied to each parameter wise into dependency. Each parameter and class-wise probability of low condition:

The total average of rule 1: Completely, 30 parameters of the diabetes dataset and 2 classes of the data. Applied each rule wise, probability of low condition (total 60 rules) is 92.63%.

Rule 2: Medium condition

If (age between 0 and 80) and (gender is M or FM) and (marital status married or unmarried) and (weight) and (height) and BMI and HbA1c and cholesterol and amylase and creatinine and insulin test and urine test and skin flood and thyroid and triglycerides and fasting sugar and 1-hour sugar and 2-hour sugar and random sugar and PP blood sugar or then medium diabetes (prediabetes of (like as stage 2)).

Age (0–80 and above) and gender (male and female is 1 and 0) and material status (material is married and unmarried is 1 and 0) weight (50–80 kg) and height (male is 1.524–1.8288 and female is 1.4224–1.6460) and BMI (male is 18.6–24.9 and female is 18.6–22.9) and FBG (100–125) and HDL (50–59) and LDL (130–159) and cholesterol (200–239) and triglycerides 150–199 FG, 1-hour, and 2-hour glucose test in 100, 75, and 50 grams (different values between 60–100 and 200–140) and 2-hour value (between 140–200) random glucose test (140–200) and creatinine 0.7–1.3 (between children's, youth, and old persons) and blood urea (normal levels depending on age (1.8–7.1 or 6–20)) and amylase-blood (23–85 in lab ranges is 40–140) and DPF (0.4–0.8) and pregnant (yes or no (yes medium 3 or 4 time)) and skin flood (30–35%) and insulin test (yes or no, medium is steady) and thyroid test (yes or no) and HbA1c (medium is 7.0–8.0) and uric acid (normal values of men is 4.0–8.5 and women 2.5–7.5) and urine test: yes and no DPF <0.4 and BP (120/80 to 140/90) and thyroid (yes or no) and skin—flood (ranges is for all 30–35%) and smoking (yes or no) and depression and stress (yes or no) and urine test (yes or no))

or then

Class (prediabetes is positive and negative)

Rule 2 is the false-positive value

In this case, each parameter wise applied into dependency. Each parameter and class wise of the probability of medium condition of prediabetes.

The total average of rule 2: completely 30 parameters of the diabetes dataset and 2 classes of the data. Applied each rule wise, probability of low condition (total 60 rules) is 85.72%.

Rule 3: High condition

If (age between 0 and 80) and (gender is M or FM) and (marital status married or unmarried) and (weight) and (height) and BMI and HbA1c and cholesterol and amylase and creatinine and insulin test and urine test and skin flood and thyroid and triglycerides and fasting sugar and 1-hour sugar and 2-hour sugar and random sugar and PP blood sugar or then high or very diabetes (diabetes of (like as stage 3)).

If (age (0–80 and above) and gender (male and female is 1 and 0) and material status (material is married and unmarried is 1 and 0) and Weight (50–80 kg) and height (male is 1.524–1.8288 and female is 1.4224 to 1.6460) and BMI (male (25–29.9 and very high is 30 and above) female (23–29.9 and very high 30 and above)) and FBG (126 and above) and HDL (60 and above (>1.33)) and LDL (160–199 (4.1–4.9)) and cholesterol (1240 and above) and triglycerides (200 and 499) and triglycerides (very high is 500 and above) and FG, 1-hour and 2-hour glucose test in 100, 75, and 50 g (different values between 60–100, 200, and 140) and 2-hour value (between 200 and above) and random glucose test (above 200 (>200)) and creatinine 0.7–1.3 (between children's, youth, and old persons) and blood urea (normal levels depending on age (1.8–7.1 or 6–20)) and amylase-blood (range 23–85 in lab ranges is 40–140) and DPF > 0.8 and pregnant (yes or no (yes high 4 or above times)) and skin flood ((body fact calculation) 30–35%) and insulin test ((yes or no) medium is steady) and thyroid test (yes or no) and HbA1c (high 9.0–14.0) and uric acid is (normal values of men is 4.0–8.5 and women 2.5 to 7.5) and blood pressure (140/90 above is higher) and smoking (yes and no) and depression and stress (yes and no) and urine test (yes and no)

or

Then diabetes (positive and negative) rule 3 is false negative.

In this case, each parameter wise applied into dependency. Each parameter and class wise of the probability of low condition of diabetes.

5.4.2 CLUSTERING ALGORITHM: BDNF GENE WITH T2DM DATA USING CLUSTERING ALGORITHM

Clustering algorithm: BDNF gene with T2DM data using clustering algorithm

Input: Patient data are initializing as matrix with each query wise

Output: Generate clusters and build dimensionality reduction matrix.

After completion fuzzy rules designing part of rules generating based on design clusters with BDNF gene data.[13,16–18]

Step 1: Start the process

Step 2: Apply posterior probability of each rule wise complete BDNF of diabetes data. Considering data each rule wise.

$$Q \text{ is row wise—}\mathbf{M}_{1*m} \tag{5.1}$$

$$R \text{ is column wise—}\mathbf{N}_{1*n} \tag{5.2}$$

So that each query or rule is generated with fuzzy. Now each query of pattern wise value is considered as row and column (attribute and record) value of the pattern.

$$\text{Matrix is } [\mathbf{MN}]_{m*n} \tag{5.3}$$

Step 3: Now compute probabilities of each pattern wise. Each query-wise Input matrix is IM

$$IM = \begin{cases} 1, T(j,1) \textit{ is } Zd \\ 0, T(j,1) \textit{ is not } Zd \end{cases} \tag{5.4}$$

Step 4: Apply proposed similarity measure in step 4, step 5, and step 6.

$$T(\mu,\gamma) = \frac{2+1\,GE\,(\mu / Y)}{G(\mu)+G(Y)} \tag{5.5}$$

Step 5: Compute probabilities of $G(\mu)$ and $G(\gamma)$.

$$G(\mu) = P(\mu)\log P(\mu) \tag{5.6}$$

$$G(\gamma) = P(\gamma)\log \mathrm{P}(\gamma) \tag{5.7}$$

Step 6: Now computed as information gain based on entropy is followed:

$$IGE\ (\frac{\mu}{Y}) = P(\mu)\cdot P(\frac{\mu}{Y}) \tag{5.8}$$

and followed to

$$G(\frac{\mu}{Y}) = -\Sigma P(\frac{\mu}{Y}) \cdot \log_2 P(\frac{\mu}{Y}) \tag{5.9}$$

Step 7: Proposed similarity measure is followed as step 4 and step 5 based on computed step 3 followed by the equations are eq 5.6–5.9 bead on eq 5.5 designed.

Step 8: Apply each and every rule or query with measure generating clustering.

Step 9: Repeat step 3 to step 5.

Step 10: Stop the clustering procedure.

5.4.3 CLASSIFICATION ALGORITHM OF k-NEAREST NEIGHBOR (KNN) ALGORITHM OF BDNF GENE WITH TYPE-2 DIABETES DATA

Classification algorithm of *k*-nearest neighbor (kNN) algorithm of BDNF with type-2 diabetes data

Input: Dimensionality reduction with rule wise.

Output: Identifying patients and severity patients.

BDNF with type-2 diabetes data using classification algorithm is kNN which is based on classified patients positive and negative dataset.[6–17]

Step 1: After applying fuzzy rules and clustering with proposed similarity measure based on generated clusters defined reduced the matrix.

Step 2: Now applied dimensionality reduction of the reduced matrix, applied classification algorithm is kNN.

Step 3: kNN classification algorithm of similarity measure and proposed similarity measure, apply two samples of each pattern and compute similarity.

Step 4: Similarity measure is existing the Euclidian distance misconstrue measure, normalized Euclidian distance measure and proposed similarity measure is Modified Sigmoid function-1 and Modified sigmoid function-2 based on compute similarity.

Most popular similarity measure is Euclidean distance measure, based on two samples of the data:

$$K(x), K(x-1), \ldots, K(x\text{–}Lx+1) \tag{5.10}$$

training vector is *i*th vector:

$$K(i), K(i-1), \ldots, K(i - Lm + 1) \tag{5.11}$$

Existing equation is Euclidean distance measure.

$$Ed(X) = \sqrt{\sum_{j=0}^{Lm-1} (K(x-j) - L(i-j))^2} \tag{5.12}$$

U and V are two coordinate values:

$$NH(Wi, Wj) = \frac{Ne(Ui, Uj) + Ne(Vi + Vj)}{2} \tag{5.13}$$

$Ne(V1, V2)$, normalized Euclidean distance between U and V:

$$\text{Ne}(Ui, Uj) = \frac{Ed(Ui, Uj)}{\max(Ed)} \tag{5.14}$$

$$\text{Ne}(Vi, Vj) = \frac{Ed\ (Vi, Vj)}{\max(Ed)}$$

Whereas equation defined as eq 5.12 followed to $Ne(Ui,Uj)$ Sigmoid measure is

$$S(x) = \frac{1}{1+e^{-(x)}} \tag{5.15}$$

Proposed Similarity Measure

Sigmoid function with Euclidian space and normalized Euclidian distance with space followed computed similarity each pattern wise is followed to proposed similarity measure

$$\text{SE (space)} = \frac{1}{1+e^{-(Ed(x))}} \tag{5.16}$$

SE (sigmoid function with Euclidian space)

$$\text{SNME (space)} = \frac{1}{1+e^{-(Ne)}} \tag{5.17}$$

SE (sigmoid function with normalized Euclidian space)

Existing (Euclidian distance measure and sigmoid function) and proposed similarity measure (normalized Euclidian, sigmoid with Euclidian space, and sigmoid with normalized Euclidean distance) is applied into each and computed pattern wise as similarity.

Step 5: After computed similarity measure and computed cutoff based on defined conditions, cutoff value is

$$Cv = \frac{Np + Nq}{Np + Nq + ANq + ANq}$$

Np = positive patients.
Nq = negative patients.
ANp = average positive patients.
ANq = average negative patients.
Cv = cutoff value.

Step 6: Conditions of the BDNF with type-2 diabetes of patient define positive and negative. Condition 1 is similarity value ≤ cutoff value = positive.
Condition 2 is similarity value > cutoff value = negative.

Step 7: Then, identify the patient's behavior and severity of the patients. The scale of severity of the patients is low, medium, high, and very high.

Step 8: Repeat step 4 to step 7.

Step 9: End of the process.

5.5 DISCUSSION AND FUTURE WORK

The designed framework GROCD of this research work is based on the prediction of patients. One of the most important challenges is the patients' behavior identification at a scale of low, medium, and high. Few challenges in this framework give solutions such as reduced data and remove noise as well. Proposed similarity measure worked and accurately predicted patients with framework, and feature work defined novel fuzzy membership based on defined accurate rules and efficient clusters.

5.6 CONCLUSION

This research work addresses one of the best challenges predicted patients and behavior of the patients. We design framework GROCD which addressed novel fuzzy rules with efficient clustering and classification of patients. This framework proposed work control ambiguity of especially reducing matrix. This proposed framework gives accurate results, followed by designed novel fuzzy rules with efficient clusters with proposed similarity measure and classification of kNN of classified patients using proposed sigmoid similarity measure with Euclidean space predicted in patients.

KEYWORDS

- **BDNF gene**
- **type-2 diabetes mellitus**
- **fuzzy rules**
- **clustering algorithm**
- **classification algorithm**
- **kNN**
- **similarity measure**
- **information gain**

REFERENCES

1. Korley, F. K.; Arrastia, R. D.; Wu, A. H. B.; Yue, J. K.; Manley, G. T.; Sair, H. I.; Eyk, J. V.; Everett, A. D.; Okonkwo, D. O.; Valadka, A. B.; Gordon, W. A.; Maas, A. I. R.; Mukherjee, P.; Yuh, E. L.; Lingsma, H. F.; Puccio, A. M.; Schnyer, D. M. Circulating Brain Derived Neurotrophic Factor (BDNF) Has Diagnostic and Prognostic Value in Traumatic Brain Injury. *J. Neurotrauma* **2016,** *33* (2), 215–225.
2. Lee, T.; Fu, C.-P.; Lee, W.-J.; Liang, K.-W. Brain-Derived Neurotrophic Factor, But Not Body Weight, Correlated with a Reduction in Depression Scale Scores in Men with Metabolic Syndrome: A Prospective Weight-Reduction Study. *Diabetol. Metab. Syndrome* **2014,** *6*, 18.
3. Patil, B. M.; Joshi, R. C.; Toshniwal, D. Hybrid Prediction Model for Type-2 Diabetic Patients. *Expert Syst. Appl.* **2010,** 8102–8108.
4. Devarapalli, D.; Apparao, A.; Kumar, A.; Sridhar, G. R. A Novel Analysis of Diabetes Mellitus by Using Expert System Based on Brain Derived Neurotrophic Factor (BDNF) Levels. *Helix* **2013,** *1*, 251–256.
5. Devarapalli, D.; Apparao, A.; Kumar, A.; Sridhar, G. R. A Multi-Layer Perceptron (MLP) Neural Network Based Diagnosis of Diabetes using Brain Derived Neurotrophic Factor (BDNF) Levels. *Int. J. Adv. Comput.* **2012,** *35* (12), 422. ISSN: 2051-0845.
6. Varma, K. V. S. R. P.; Rao, A. A.; Lakshmi, T. S. M.; Rao, P. V. N. A Computational Intelligence Approach for a Better Diagnosis of Diabetic Patients. *Comput. Electr. Eng.* **2014,** *40*, 1758–1765.

7. Varma, K. V. S. R. P.; Rao, A. A.; Lakshmi, T. S. M.; Rao, P.V. N. A Computational Intelligence Technique for the Effective Diagnosis of Diabetic Patients Using Principal Component Analysis (PCA) and Modified Fuzzy SLIQ Decision Tree Approach. *Appl. Soft Comput.* **2016,** 137–145.
8. Canedo, V. B.; Marono, N. S.; Betanzos, A. A. A Review of Feature Selection Methods on Synthetic Data. *Knowl. Inf. Syst.* **2013,** *34*, 483–519. DOI:10.1007/s10115-012-0487-8.
9. Zeki, T. S.; Malakooti, M. V.; Ataeipoor, Y.; Tabibi, S. T. An Expert System for Diabetes Diagnosis. *Am. Acad. Scholarly Res. J.* **2012,** *4*, 1.
10. Chen, W.; Song, Y.; Bai, H.; Lin, C.; Chang, E. Parallel Spectral Clustering in Distributed Systems. *IEEE Trans. Pattern Anal. Mach. Intell.* **2011,** *33* (3), 568–586.
11. Kramer, O. *Dimensionality Reduction with Supervised Nearest Neighbors*; Springer, Intelligent Systems References Library, 2013; vol 51.
12. Jiang, S.; Pang, G.; Wu, M.; Kuang, L. An Improved *K*-Nearest-Neighbor Algorithm for Text Categorization. *Expert Syst. Appl.* **2012,** *39*, 1503–1509.
13. Ghoting, A.; Parthasarathy, S.; Otey, M. E. Fast Mining of Distance-Based Outliers in High-Dimensional Datasets. *Data Min. Knowl. Discov.* **2008,** *16*, 349–364.
14. Dessi, N.; Pes, B. Similarity of Feature Selection Methods: An Empirical Study across Data Intensive Classification Tasks. *Expert Syst. Appl.* **2015,** *42* (10), 4632–4642.
15. Kumar, V.; Minz, S. Multi-View Ensemble Learning: An Optimal Feature Set Partitioning for High-Dimensional Data Classification. *Knowl. Inf. Syst.* **2016,** *49*, 1–59.
16. Xu, R.-F.; Lee, S.-J. Dimensionality Reduction by Feature Clustering for Regression Problems. *Inf. Sci.* **2015,** *299*, 42–57.
17. Lee, S.-J. Multi Label Text Categorization Based on Fuzzy Relevance Clustering. *IEEE Trans. Fuzzy Syst.* **2014,** *22* (6), 1457–1471.
18. Panigrahi, S.; Rajasekhar, N. A Novel Cluster Evolution for Gene-miRNA Interactions Documents Using Improved Similarity Measure. In *International Conferences on Engineering & MIC—2016 (ICEMIS-2016)*, IEEE Morocco Section, IEEE, 2016; pp 1–7.
19. Mangathayaru, N.; Mathurabai, B.; Srikanth, P. Clustering and Classification of Effective Diabetes Diagnosis: Computational Intelligence Techniques Using PCA with KNN. In *Springer International Publishing AG 2018 Information and Communication Technology for Intelligent Systems (ICTIS 2017) me 1, Smart Innovation, Systems and Technologies*, SIST Vol 83, 2018; pp 426–440.
20. Srikanth, P.; Deverapalli, D. A Novel Cluster Algorithms of Analysis and Predict for Brain Derived Neurotrophic Factor (BDNF) Using Diabetes Patients. In *International Conference on Computer and Communication Technologies (IC3T 2016), Springer 2016, Advance Intelligence System and Computing (AISC)*; Vol 542, 2016; pp 109–125.

CHAPTER 6

Hybridization Preprocessing and Resampling Technique-Based Neural Network Approach for Credit Card Fraud Detection

BRIGHT KESWANI[1*], POONAM KESWANI[2], PRITY VIJAY[3], and AMBARISH G. MOHAPATRA[4]

[1]*Department of Computer Applications, Suresh Gyan Vihar University, Jaipur, Rajasthan, India*

[2]*Akashdeep PG College, Jaipur, Rajasthan, India*

[3]*Suresh Gyan Vihar University, Jaipur, Rajasthan, India*

[4]*Electronics and Instrumentation Engineering, Silicon Institute of Technology, Bhubaneswar, Odisha, India*

[*]*Corresponding author. E-mail: kbright@rediffmail.com*

ABSTRACT

Credit card fraud is a form of financial fraud growing every year, causing losses to financial multinationals as well as government sectors. Traditional methods like manual detection of credit card frauds are feasible only for small datasets, but with the rise of big data, these methods are of no worth. Data mining and machine learning (ML) techniques have been widely used for fraud detection, by drawing a pattern that separates two classes (fraud and legitimate). The only necessity of ML algorithms is vast sets of examples, but messy and complex datasets are tricky for ML techniques and thus bring challenges during formulating learning rules. Almost all real-world dataset is unclean which tends to produce inaccurate models. This chapter revealed various challenges possessed by ML classification algorithms because of complex imbalance dataset and therefore proposes a new hybridization preprocessing and resampling technique (HPRT), which solves two major issues: (a) cleans the dataset while (b) balancing the dataset. HPRT thus enhances the performance of ML algorithms. Several ML algorithms were observed to produce better result when combined with HPRT. HPRT-based neural network model is constructed for detection of credit card fraud and is compared with traditional neural network model. Confusion matrix and other matrices revealed moderately high and precise results for HPRT-based neural network-based model in comparison to traditional neural network model.

6.1 INTRODUCTION

Dataset state of imbalance occurs when more numbers of examples for one class are present in it. A class having examples more in number are called majority class and other is called minority class. This problem is not new during mining the data. Since past years, the researchers are continuously looking toward finding a solution for learning from imbalanced data but still, this issue is an important topic for research. With the alighting of big data, both data mining (DM) and machine learning (ML) technology overemphasizing to grow into one of the extensive tools for eradicating deep insight from imbalance dataset. Dataset imbalance problem, on the other hand, gained lots of importance with the advancement of big data[1] and with it also bought much threat while gathering valuable information from it.[2] Upgradation of data level and algorithm level methods have been noticed with many new hybrid approaches.[3] Many domains do not have a balanced dataset. Minority class consists of the most important information and extracting that piece of information from the big dataset is an actual challenge. During the processing of these datasets, minority class does not get importance and therefore classification shows inaccurate results. These real-life imbalance dataset challenges galvanize researchers and scholars to focus on an effective and real-time solution for such a problem. Fraudulent credit card transactions, intrusion detection, hardware fault detection, insurance risk modeling, and so on are some of the examples of real-world imbalance dataset.

Before the classification of imbalance complex datasets, resampling methods are headed for balancing such dataset. Old resampling methods like oversampling (OS) and undersampling (US) techniques balance the dataset but have lots of disadvantages. OS when applied can cause problem of overfitting and US cause's information loss. Hybridization preprocessing and resampling technique (HPRT), an algorithm which will automatically balance the dataset without much information loss, and at the same time, the classifier will not cause any overfitting problem. Our method uses both US and OS, where US occurs at majority class through reduction of redundancy and extreme outliers. This step can reduce majority class, almost up to 40% and then in the next step SMOTE OS is applied to a minority class. This two-step approach can balance the dataset while cleaning it and as a result ML algorithm performance can enhance.

HPRT consists of several steps; each step is designed to reduce the complexity of the dataset automatically. This algorithm is basically designed for large and big datasets, where once an imbalance and complex dataset are input to an algorithm; HPRT will reduce the complexity of the dataset and convert it to form a balance state in several steps. ML techniques are applied in this algorithm thus the process is automated. In the first step, PCA reduces the dimensionality of a big dataset. Minority class is then separated which avoid losing any kind of information in it after which *K*-mean unsupervised clustering algorithm divides majority class into several clusters depending upon the business problem. This step divides the similar kind of data feature in the same cluster and hence minimizes comparison between features of majority class. Redundancy is reduced using divide and conquer rule within each cluster and then extreme outliers have been dropped from each cluster with the help of the Tukey method. In this way, almost 30%, 40% of the complexity had been removed. In the next step, synthetic sample is added to minority class during cross-validation to avoid data leakage problems and reduce the chances of overfitting. Once the dataset passes through all these steps of the proposed algorithm, it automatically balances dataset and then we can apply ML binary classification for construction of the predictive model. The HPRT is an enhanced and new resampling preprocessing approach for optimizing the complexity of imbalance large and big dataset through reduction of redundancy and extreme outliers from majority class and then sample is increased in minority class through SMOTE. This enhanced algorithm is applied to the credit card transaction imbalance dataset used as a sample dataset. Thus, it successfully reduces the complexity of the dataset and proved to be enhancing resampling technique for balancing of the dataset.

6.2 PCA AND *K*-MEAN CLUSTERING IN HPRT

PCA is a dimensionality reduction method which extracts the set of features from a very large or big dataset and converts it into low-dimension dataset. In other words, it extracts all the information from a high-dimensional dataset with an intention to capture almost all the information from a big and high dimensional dataset. Increased size of data is not only hazards for storing but processing it also becomes problematic for many traditional ML approach. Nowadays, sizes of data anticipating from different sources have been enlarged containing in numerous features in it.[4] Out of these numerous features, many are redundant and convey the same piece of information. Therefore, this type of redundant features should be removed so that dataset contains less but very important and meaningful information. PCA checks the variance of each property into a dataset and collects all the high variance data to form a new set of examples based on the original one while absorbing all the information in it. PCA linearly transformed high dimensional original data using the algebraic calculation of principal component.

Clustering is a technique of assembling a group of similar elements. Dataset divides into different groups by placing identical items in a similar group to form a cluster. Features in the same cluster contain the same properties as much as possible. Features in the different cluster will be different altogether. Clustering is an unsupervised technique which is used in almost every domain. Nowadays big companies like Amazon, Netflix, Facebook, and so on are using clustering to manage the data appropriately.[5] Banks are using clustering technique for analyzing customer probabilities and their credit score. Finance companies are using clustering, for credit card detection, fraud detection, and identification of risk factor. It is used in market segmentation, image segregation, and also in grouping web pages. In business problems, clustering helps to formulate rewarding decision by analyzing customer behavior for shopping.

With the augment of big data expansion in the volume of data occurs speedily. Big data is a term given to the large volume of the complex dataset processing of which is not possible using traditional methods.[7] Therefore, clustering is very important DM method used in big data analytics.

K-means clustering is an unsupervised ML approach as it is used for classification when a label is not present in the data. *K*-means algorithm explicitly accepts the value of *K*, as it divides the dataset into *K* number of groups by placing a similar item in one group. The algorithm performs a number of iterations to assign each data point to groups depending upon its feature. *K*-means define centroid for each group *K*, with the help of which new data can be labeled. It classifies the training dataset by defining labels for them. *K*-means algorithm allows analyzing the cluster which is formed systematically. The centroid is the center point of the group depending on which other data points are placed on a cluster accordingly. *K*-means algorithm is used in many domains for segregation of facts into groups. It is very useful in business problems for grouping purchase history, several activities, clustering inventories depending on various modes, finding groups for images; it also benefits in health care by clustering patient records into a different category. *K*-means are used in the detection of anomalies also. It helps to discards, noisy data, like outliers, and redundant items more efficiently by tying them together. In a dataset containing a huge amount of data, redundancy reduction or outlier's reduction and other cleanup processes are a big challenge if data are not categorized properly because a number of comparisons are required in such a situation. Therefore, to make the task easier and simpler, data should be huddled in the same group based on similarity of an item.

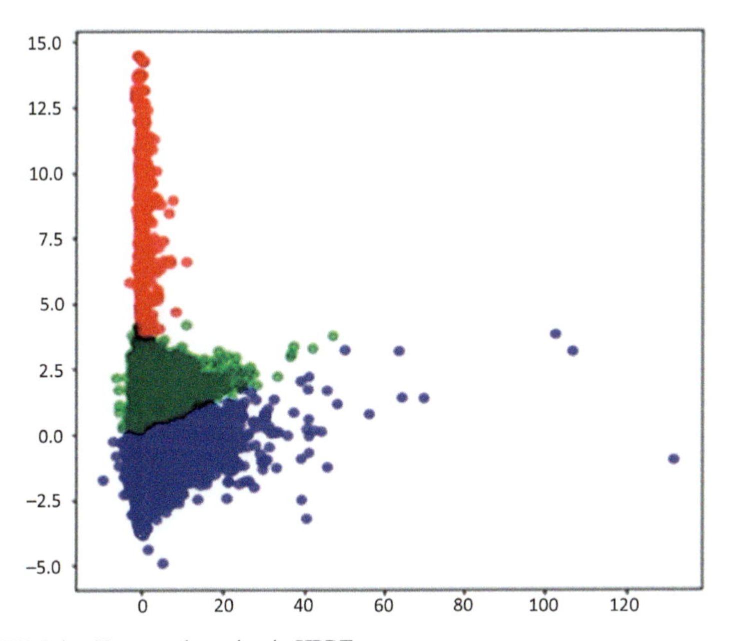

FIGURE 6.1 *K*-mean clustering in HPRT.

6.3 OUTLIER: A PROBLEM FOR ML ALGORITHMS

Outliers are extreme values in an observation of dataset that falls beyond the range from other records that are more or less similar to each other.[6] Outliers can occur in a dataset due to some experiment error variation in measurement, and so on. Outliers in a dataset should be taken care of before application of ML algorithm to the dataset to get an appropriate result because ML algorithm is sensitive toward the distribution of the dataset. A massive amount of outliers in a dataset can cause poor and inaccurate models with longer training times by misguiding ML algorithms. In a dataset, outliers can be part of records during the collection, processing, or analyzing records. ML algorithm like linear and logistic regression is very much affected in its training process with the presence of outliers. Human error, instrument error, experimental errors, data processing errors, and so on are some of the causes of outliers. Outliers can be broadly classified into two types: univariate and multivariate.

Univariate outliers occur when extreme values are searched on one feature of the dataset. Box plots fall under the category of a univariate method. It is one of the simplest and popular techniques for the detection of outliers. Box plot describes a feature of the dataset by using statistics calculation, such as lower quartiles, upper quartiles, and median. Box plot uses the format of the box for specification of data distribution. Box plot, known as Tukey's method, introduced in 1977 is a very important visualization tool for the detection of outliers by displaying univariate features as lower quartile, upper quartile, lower extreme, upper extreme, and median of the records present in the dataset. IQR is an interquartile range, which is a distance between Q1 and Q3 quartiles. Inner fences are the place at 1.5 IQR below Q1 and above Q3 distance. Outer fences are at the place of 3 IQR below Q1 and above Q3. Features detected between the inner and outer fences are the possible outliers and values beyond outer and inner fences can be identified as extreme outliers. Multivariate method occurs when extreme values are searched throughout the dataset. Multivariate method built a model for outlier detection. It uses all the data at once and solves the problem by cleaning the instances having errors above a value specified. HPRT used with the Tukey method with *K*-means clustering to detect extreme outlier. Care is taken to drop only extreme outliers without losing important information. Dropping extreme outliers present in majority class of our large dataset will help in balancing the dataset, to some extent, without any fear of vital information loss which creates positive effects on the accuracy of the model. Threshold is determined by multiplying IQR with 1.5. This value acts as a threshold for detecting extreme outliers, and therefore, not much vital information will discard from the dataset. Value beyond threshold range will consider as outliers and can be dropped from the dataset. Box plot is used for the visualization of outliers.

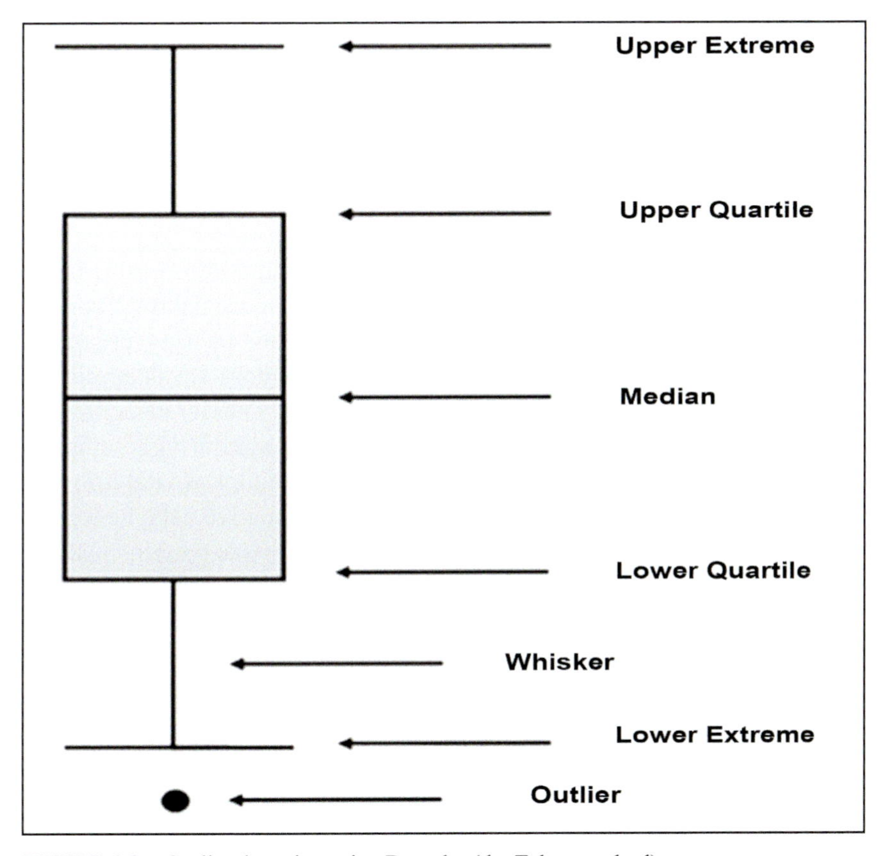

FIGURE 6.2 Outlier detection using Box plot (the Tukey method).

6.4 REDUNDANCY: A PROBLEM FOR ML ALGORITHMS

Redundancy is a phase where more than one copy of data is present in the database. The same piece of more information can cause inconsistency in a database. Data redundancy occurs either circumstantially during storage or deliberately during backup purposes.[8] Predictive model accuracy depends upon the quality of training and pattern used while learning process. Therefore, features present in a dataset play a very important role for the training of model as well as in its accuracy. In such conditions, redundant and nonrelevant features in a dataset end with the construction of a weak model resulting poor accuracy during prediction.[9] However, many researchers have been conducted to overcome this situation.[15] With the invention of big data, the clustering algorithm has been used to a great extent to check for redundant data and removes it from a dataset.

6.5 TOWARD SOLUTION: HPRT AS A TOOL TO ENHANCE ML ALGORITHMS

Two popular resampling methods, random US and SMOTE OS, cause certain challenges during its application to an imbalanced dataset. Random US, abandon the majority class features randomly causing enormous information loss. Consider the scenario, where random US is applied to big data for business decision making or medical big data for detecting any severe disease or credit card fraud detection dataset and during resampling features reduced randomly can discard immense amount of vital information and leave the dataset with not as much of informative data; hence, the traditional ML algorithm cannot learn a strong predictive pattern. SMOTE cannot perform well with high dimensional big data, and application of it during resampling big data will cause several complications. Therefore, the goal of this research is to develop an enhanced and hybrid ML resampling approach, which can act as a perfect solution for reducing the complexity of big data because of its imbalance nature.

With the rise in big data complexities of the datasets also increases, thus traditional methods like random US, random OS, OS through SMOTE suffers from various challenges. Therefore, intention is to propose enhanced ML approach and masquerade the limitation of traditional resampling methods. Here in this enhanced hybrid resampling approach, both US and OS are applied to a dataset for balancing it but in a different approach. In the first step, feature reduction of the majority class is done by removing extreme outliers and redundant data. Redundancy in a dataset occurs when a piece of information exists for more than one time, either intentionally or unintentionally. Redundancy is consisting of repetition the variable, the existence of which brings inconsistency in dataset. Outliers are the extreme values that do not match with other observations in a dataset. Occurrences of redundant as well as outlier's data in a dataset will make it noisy and removing it will not generate much information loss. In the second step, minority class is increased through adding data point synthetically to it through SMOTE. Keeping this in mind, HPRT is proposed, which is an enhanced automated ML-based resampling approach, which can automatically convert the imbalance and complex structure of a dataset. Our enhanced technique automates the process of resampling for big datasets by detecting the imbalanced nature of a dataset and converting it into balanced dataset automatically.

HPRT accepts original imbalance dataset as an input. The dataset than undergoes through several steps, wherein each step complexity of the large dataset has been tried to reduce. In the first step, an algorithm accepts big or large dataset as an input and then divides it into two subparts, such that first part contains all the features from majority class and another part contains all the features from minority class. In this way, two subsets are derived from the original dataset are N and F where N consists only majority features and F consists minority features. In the next step, we individually apply PCA algorithm on both the subset to form P1 and P2. This is a first phase, where the complexity of the dataset is reduced through the dimensionality reduction. In the next step, we will take only subset P1{V1, V2, V3, ... V*n*} and divide it into various clusters with the help of *K*-mean unsupervised ML algorithms. *K*-means is capable of dividing observations in datasets into *K* different clusters. Clusters can be defined as the entities of a similar group, that is, features in one cluster are identical that the features in the other clusters. With the help of *K*-mean clustering algorithm P1{V1, V2, V3, ... V*n*} is divided into P1{k*i*, k*j*, ... k*n*}. The cluster is developed to tie similar items in one group so that the number of comparisons will decrease to an enormous extent and which is helpful in reducing the processing time of big and large dataset. Redundancy is checked in each cluster using divide and conquer rule. Duplicate record detected is immediately discarded from the cluster.

After the removal of unwanted redundant data from the clusters P1{k*i*, k*j*, ... k*n*}, the algorithm proceeds with the detection of outliers. IQR method is used for the detection of extreme outliers. The threshold is calculated by multiply IQR with 1.5; upper bound and lower bound are also calculated.

A feature is considered an outlier if its value is less than lower bound or more than the upper bound. All the outliers are stored in temporary list O and discarded at last from the cluster. At this stage, our clusters P1{V1, V2, ... V*n*} are free from outliers as well as duplicates data and then the cluster is again converted into a data frame. Half of our intension to reduce the majority class is achieved through these steps. Hence, this reduction in the majority class is achieved without any type of information loss. In the next step, we concatenate both subset P1{V1, V2, V3, ... V*n*} and P2{V1, V2, V3, ... V*n*} to form D{V1, V2, V3, ... V*n*}. Then, in the next step, SMOTE OS is used to balance minority class of the dataset. In this process, features in minority class are increased by adding artificial sample to the minority class with the help of the KNN algorithm and Euclidean distance. SMOTE OS is performed during cross-validation to avoid overfitting because, if applied earlier, there can be a possibility of adding features exactly the same in the validation set, hence causing the problem of data leakage. To avoid this, validation set must be excluded first with other training set and then during cross validation OS should be applied on rest of the dataset. This will avoid overfitting as well as data leakage problem because during the testing period validation set will be completely unseen.

Above technique balances the dataset thus reducing its complexity and then in next part binary classification algorithm,[14] such as logistic regression, *K*-nearest neighbor, and decision tree had been applied on the balanced dataset, so as to construct credit card fraud detection models. The models are evaluated with the help of a confusion matrix, sensitivity, precision, accuracy, and misclassification. Confusion matrix will give a total count of true negative (TN), true positive (TP), false negative (FN), and false positive (FP). TN is a correct count for negative class, TP is a correct count of positive class, FN is a wrong prediction for negative class, and FP is a wrongly classified positive class.

Logistic regression predicts the result based on probabilities and sigmoid curve value. The value of sigmoid curve ranges from 0 to 1. Probabilities calculated from the input feature during logistic regression process pass through the sigmoid curve. If the output is greater than or equal to 0.5, then the class is predicted as 1 (positive class) and if the output is less than 1 then the class is predicted as 0 (negative class).

KNN predicts the result after calculating the distance of unknown features with that of given features distance. Euclidean distance is used for distance calculation here. The result is predicted based on a number of

K-nearest neighbor which is provided by the user. In this study, parameter for *k* is 5, that is, the algorithm will consider five nearest data points with that of the unidentified feature, and then output class for that feature will be calculated by counting class of the data points appears for a maximum number of time.

Decision tree builds models by defining rules for training examples. During this process, trees are constructed to solve the classification problem. Generation of the tree starts with a selection of root node from the training dataset and then the process continues to find branches, subbranches, and so on. The process of searching roots and its subset is repeated for each branch of the tree. Gini index is a matrix used in this experiment for selection of roots during tree construction. It identifies root node by measuring how often attribute chosen randomly can be incorrect.

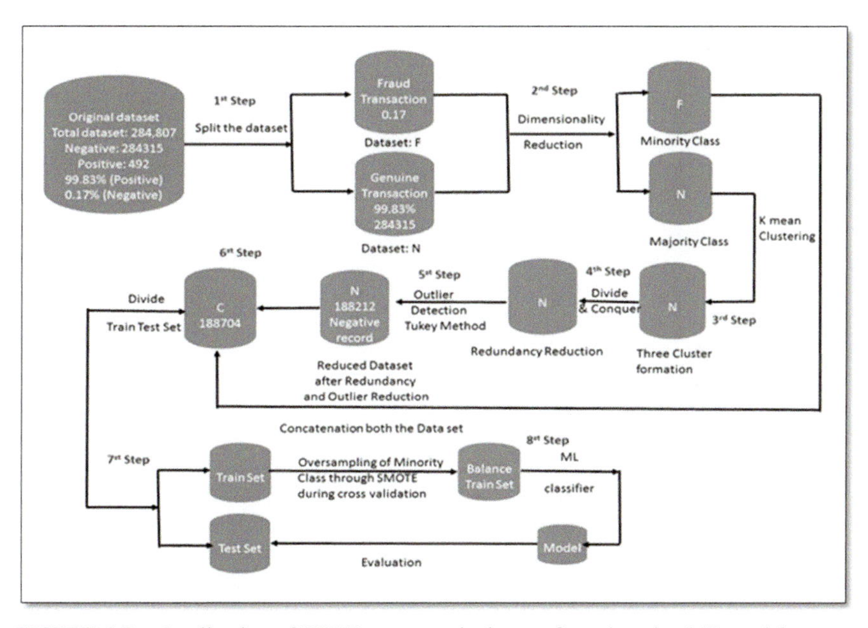

FIGURE 6.3 Application of HPRT on a sample dataset for enhancing ML models.

6.6 DESIGNING OF HPRT TO ENHANCE ML ALGORITHMS

A new tool HPRT is used for optimizing complexity of an imbalanced big data. It is an enhanced resampling and preprocessing technique for balancing the dataset while cleaning it, such that enhancing performance of ML classifier.

Input: Imbalanced Big Dataset C $\{V_1, V_2, V_3, \ldots Vn\}$
Output: HPRT-based ML Model constructed on top of balanced dataset

Step 1: Load dataset C, divide it into two parts (N, F)

F consist only Minority Class

$$F \in C$$

N consist only Majority Class

$$N \in C$$

Step 2: Reduce the dimensionality of both N and F separately through the application of PCA:

1. Standardize dataset N and F so that values of all the features V1, V2, V3 … V*n* exists within 0 to 1 without changing its original meaning.

$$N = \text{Standardize}\ (N\{V1, V2, V3, \ldots Vn\})$$
$$F = \text{Standardize}\ (F\{V1, V2, V3, \ldots Vn\})$$

2. Calculate covariance matrix for given set of data:

$$V1 = \text{Cov}(N\{\{V1, V2, V3, \ldots Vn\})$$

$$V2 = \text{Cov}(F\{\{V1, V2, V3, \ldots Vn\})$$

3. Calculate Eigen values and Eigen vectors for matrix V1 and V2

$$E1, E2 = \text{Eigen}\ (V1)$$
$$E3, E4 = \text{Eigen}\ (V2)$$

Z1 = Select K1, Eigen values having *K*, largest Eigen values (E1, E2)

Z2 = Select K2, Eigen values having *K*, largest Eigen values (E3, E4)

4. Project data by multiplying the original matrix with its transpose.

$$P1 = Z1T \cdot N$$
$$P2 = Z2T \cdot N$$

where,
P1= N{V1, V2, V3, … V*n*}
P2 = F {V1, V2, V3, … V*n*} undergone through PCA process.

Step 3: Apply *K*-means clustering on P_1 to divide it into k*n* number of cluster.

1. P1 = P1{k1, k2, … k*n*}
2. Select centroid for P1{C1, C2, … C*n*} for each of the cluster P1{k1, k2, … k*n*} in P1.
3. Calculate the distance between each vector in a cluster P1{k1, k2, … k*n*} and search for the closest centroid.
4. Evaluate new centroid for each cluster k1, k2, … k*n* applying divide and conquer rule.
5. Repeat above process 2, 3, 4 till centroid values for P1{k1, k2, … k*n*} becomes constant.

Step 4: Dropping redundant data from each of the clusters in P1{k1, k2, … k*n*}

1. Comparison for detection of redundant data points within cluster P1{k1, k2, … k*n*}

For i = 1 to k*n*

For j = 1 to number of data points X in k*i*
Search for redundant records using divide and conquer rule and store in R.
R = duplicate features X*i*
Discard R form the clusters
End For

End For

Note: P1{k1, k2, … k*n*} after redundancy reduction is left from which outliers are removed in the next step.

Step 5: Dropping extreme outliers feature X from each cluster P1{k1, k2, … k*n*} using IQR

1. Sort each cluster P1{k1, k2, … k*n*} in ascending order
2. Calculate interquartile range within each of cluster P1{k1{X1, X2, X3, … X*n*}, k2{X1, X2, X3, … X*n*} … k*n* { X1, X2, X3, … X*n*}}

For i = 1 to k*n*

For j = 1 to X*n* number of data point

Note: Calculate 25 %(q1) and 75% (q2) of data in each cluster k*i* within for loop

IQR = q2 – q1

Threshold = IQR * 1.5

Note: Calculating upper bound (u) and lower bound (l) for each data point X*j* in cluster k*i*

l = q1 – threshold

u = q2 – threshold

End For

For j = 1 to X*n* number of data point

If X*j* < l or X*j* > u

Note: Store extreme outliers X*j* in O

O = X*j*

Drop O

End For

End For

Note: P1{k1, k2, … k*n*} is free from outliers at this stage. Good amount of redundant and outliers feature is reduced from P1{k*i*, k*j*, … k*n*} without losing much information.

Step 6: P1{k*i*, k*j*, … k*n*} is converted to data frame again to form P1{V1, V2, …, V*n*}. D is dataset after concatenation of P1{V1, V2, V3, … V*n*} and P2{V1, V2, V3, …, V*n*}

$$D \,\Xi\, P1 + P2$$

Note: SMOTE OS is applied to minority class in the next step

Step 7: Split D{V1, V2, V3, …, V*n*} into X1 and Y1 such that X1 consists an entire set of data excluding target class and Y1 consists target class only.

X1 = D. Drop {target class}

X2 = D. Drop {target class}

Step 8: Generate train and test set splits randomly using stratified split with number of splits = 10

S = StarifiedShuffle (Split = 10, test-size = 0.2)

For train-index, test-index in S (X1, Y1)

$$X_{Train'}, Y_{Train} = X_1 \text{ [train-index]}, X_1 \text{ [test-Index]}$$

$$X_{Test'}, Y_{Test} = Y_1 \text{ [train-index]}, Y_1 \text{ [test-Index]}$$

Note: XTrain, YTrain is 80% of data from dataset used for training a model XTest, YTest is 20% of data from the dataset for testing model prediction after training. The above set of algorithm will create 10 sets of XTrain, YTrain, XTest, YTestfor validation purpose where in each iteration, from 10 sets 1 will act as test set and other nine will remain as training set.

Step 9: Using GridSearchCV for searching best parameters for the classifiers.

Step 10: Using SMOTE during cross-validation for adding synthetic features in minority class

For train in split XTrain, YTrain

P = SMOTE (XTrain, YTrain)

Step 11: HPRT-based ML model for classification future data

M = P.Classifier (XTrain,[train] ,YTrain[train])

Note: M is a model constructed using P (HPRT applicant subset) with ML classifier

6.7 EXPERIMENT I: COMPARISON OF HPRT-BASED SELECTED ML CLASSIFIER

HPRT is applied to a highly imbalanced credit card transaction dataset. It consists of two-day transaction of European cardholders having two classes: fraud (0.17%) and nonfraud (99.83%). HPRT can be applied for large and big dataset as a preprocessing algorithm for reducing the complexity of an imbalanced dataset. This algorithm automatically detects the level of imbalance in a dataset and then reduces it automatically in various steps. Then three different ML classifiers (LR, KNN, and DT) are used to construct a model for the balanced dataset. Confusion matrix and various other measures evaluate the model which shows KNN as the best model with (99%) accuracy, precision (0.48%), Recall (0.91%), $F1$ score (0.58%), and misclassification (0.014%). Confusion matrix confirms that out of 37,741 test data only 55 times model predicted wrong result. Performance of decision tree is also satisfactory with 64 wrong predictions having accuracy (99%), precision (35%), recall (92%), $F1$ score (48%), and misclassification (0.016%). Logistic regression performs worst among them, having an accuracy (98%), precision (11%), recall (92%), *F*1 score (21%), and misclassification (0.018%) which is higher can that of other two model. Confusion matrix shows 689 wrong predictions among 37,741 test data. All the models constructed during this study prove application of proposed as a preprocessing step enhances their performances in terms of predicting frauds and nonfrauds accurately. KNN fails to construct good models with dataset having high volume and high dimensionality. Big data computation cost is very high with KNN algorithm, as the distance is calculated for every data points present in the dataset. Therefore, the model constructed by the KNN algorithm is not at all suitable for big data. HPRT-based robust ML algorithm can be a perfect combination for optimizing imbalance and complex big data. Therefore, HPRT-based neural network (NN) is used in the next section for big data classification.

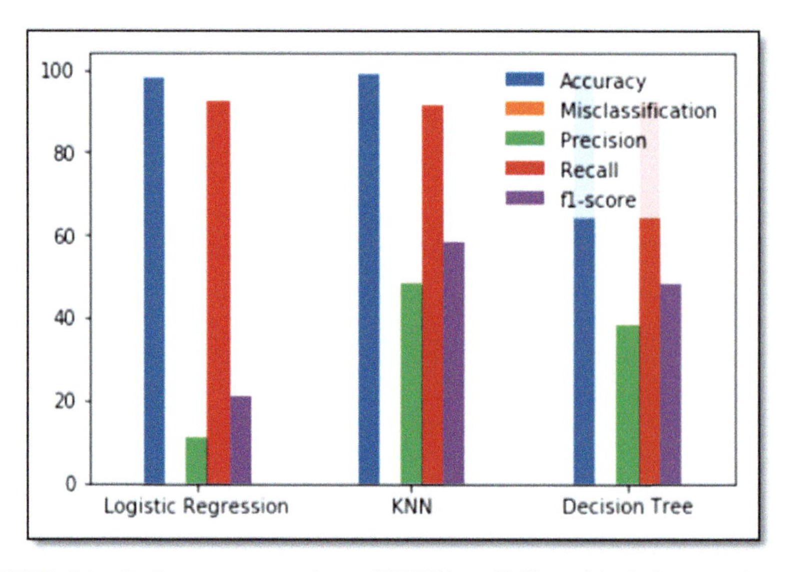

FIGURE 6.4 Performance comparison of HPRT-based ML models during experiment I.

6.8 EFFECTUAL CLASSIFICATION OF BIG DATA WITH AN HPRT-BASED NEURAL NETWORK

NN is a robust ML algorithm for classification of datasets; here, the model is built by conjoining NN with an enhanced integrated ML algorithm,

TABLE 6.1 Results of HPRT-Based ML Models during Experiment I.

Matrices and Measures			L R		KNN		DT	
	Accuracy		0.98%		0.99%		0.99%	
	Precision		0.11%		0.48%		0.38%	
	Recall		0.92%		0.91%		0.92%	
	F1 – Score		0.21%		0.58%		0.48%	
	Misclassification		0.018%		0.014%		0.016%	
	Confusion Metrics							
Total Text Samples 37741	Predicted Negative	Predicted Positive	36962	681	367603	4	37592	51
Actual Nagative	TN	FP	8	90	15	83	13	85
Actual Positive	FN	TP						

which boosts up the performances of the NN-based classifier for such highly imbalanced datasets. For this purpose, NN is used to build classification model, but before that, we have to optimize the complexity of the dataset with the help of an HPRT.

We conducted two experiments for constructing NN model for classification of frauds and nonfrauds from credit card transaction dataset: (a) in a first experiment an imbalanced dataset is a feed to traditional NN classifiers. The first model is consisting of imbalance dataset as input to traditional MLP architecture (one input layer, one hidden layer, and one output layer). Due to imbalance, nature of the dataset model does not perform well during classification of frauds and nonfrauds. The model achieves a high accuracy with low *F*1 score. Confusion matrix presents a lot FN, that is, many fraudulent transactions are detected as nonfraudulent. These types of model possess loss to finance industry. Traditional NN model results are accuracy (99%), recall (78%), precision (75%), and recall (39%) rate. Confusion matrix results display 71 misclassifications out of which 60 times the model predicts frauds as genuine, which is very costly for any model. Therefore, model was not further analyzed and dropped with an intention to construct an efficient MLP-based classifier. (b) In the second experiment, the NN was combined with HPRT and then applied to an imbalance dataset to make it balanced after which NN is used for building a predictive classification model. The second model result was outstanding, predicting both frauds and nonfrauds correctly with very less misclassification. Accuracy matrix shows almost 99% accuracy with good precision (100%), recall (85%), and *F*1 score (89%) rate. Confusion matrix produces a good result with very less number of misclassification (21 times).

6.9 ARTIFICIAL NEURAL NETWORK: AN INTRODUCTION

The artificial NN is an ML approach inspired by the brain processing system and helps in the giving out of a large amount of data. It acts just like human brain's neural structure. The human brain is a collection of neurons where each neuron is connected to form a huge and complex set of interconnected neurons network. These interconnected neurons behave like switches and change their state to produce an output when they receive input signals of a defined threshold. In the complex architecture of NN, where many neurons are connected together, the input of one neuron can act as the output of others and again behaves as input for many other neurons, and many neurons are activated again and again to learn different rules to produce output.

In 1943, Warren McCuloch and Walter Pits[10] published a research paper to narrate working principal of neurons for the first time. In that paper, they described an NN-based model proposed by them. The model was self-possessed with an electronic circuit that can formulate meaning and determine patterns from complex dataset not possible for the human brain. ANN does not follow the traditional prototype to solve the given problem where a set of instruction is compulsory. ANN is smart enough to figure out output for the issue stated by learning given examples on its own through experiences. ANN algorithm given a set of examples can construct a model for prediction of future data based on the previous examples, but the challenge here is to choose the training examples carefully because model depends on the quality of examples provided as input for the algorithm.

6.10 ARTIFICIAL NEURAL NETWORK: ARCHITECTURE

ANN mimics biological neurons, therefore neurons present in ANN network works just as neurons of the brain.[11] ANN neurons get activated by the activation function, once an input received exceeds a certain value and change its state to produce the output that is from 0 to 1 or 1 to 0. Many activation functions are used in a neural net. The most popular type of activation function is Sigmoid, Tanh, and rectified linear unit (RELU). ANN architecture is connected with several neurons where the output of one neuron becomes input to other. In this hierarchical and complex architecture, each network can be represented as nodes. Node is a center which takes the weighted inputs, calculates them, and sends them to the activation function.

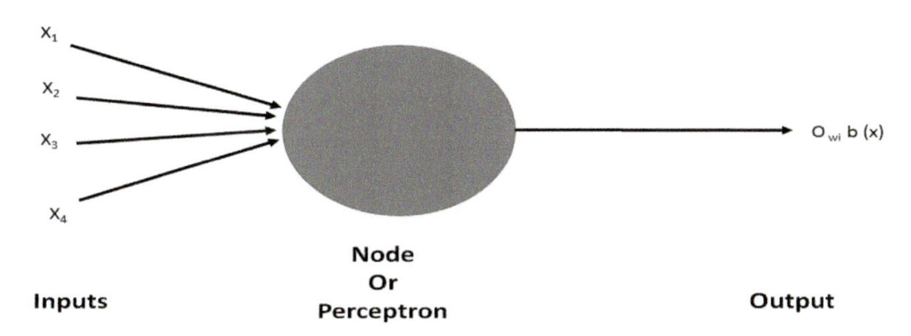

FIGURE 6.5 Node.

In Figure 6.6, W*i* is a weight in binary number multiplied by input and added up in the node as shown in eq (6.1); and *b* is a bias that helps to change the state of the output.

$$X1W1 + X2W2 + X3W3 + X4W4 + b \quad (6.1)$$

Figure 6.6 is a simple architecture of NN containing an input layer, an output layer, and a hidden layer.[11] In layer 1, each node is connected with hidden layer nodes and the hidden layer is connected to the output layer. All the connection of ANN architecture has weights associated with it. A NN receives input through the input layer; the hidden layer processes those input with the help of numerous neurons and then reflects the result to the output layer. This process is known as forwarding propagation. But every neuron contributes some error with an output which results as a variation to the desired output. To search for the seat of maximum error, neurons again travel back through the network and minimize the error by adjusting its weight. This is known as backward propagation. For minimizing, error NN uses some optimization algorithm, such as gradient descent to complete the optimization task in an efficient way.

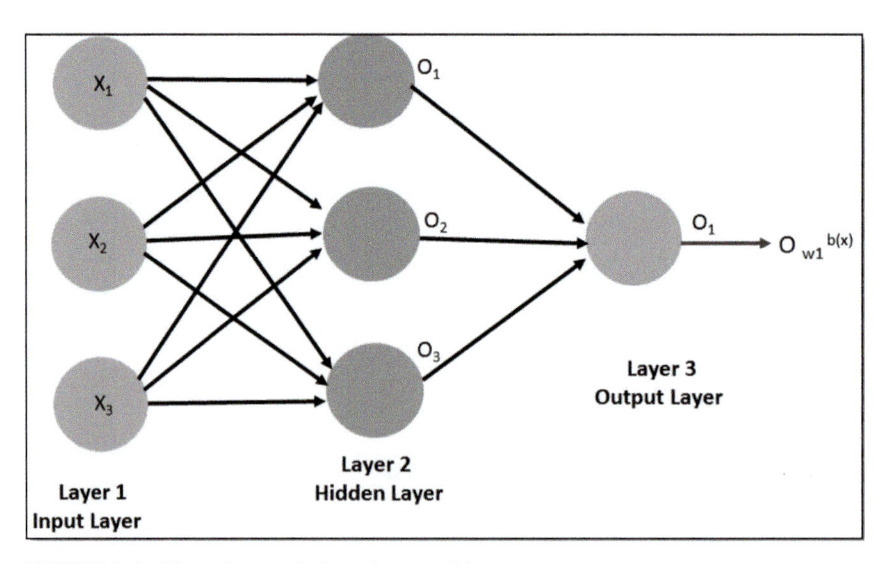

FIGURE 6.6 Neural network three-layer architecture.

Perceptron is a basic central unit of the NN responsible for the processing of numerous inputs and generates signal output. An output of a neuron is generated by calculating weighted input of all the neuron of a layer which is performed in a network through activation function. For most of the real-world complex problem, the simple NN is not sufficient. In such cases, we have to look forward to MLP which is comprised of a number of hidden layers in between input and output layer depending upon the complexity of the problem.

6.11 STOCHASTIC GRADIENT DESCENT: AN OPTIMIZER

Stochastic gradient descent, as an optimizer, is used[12] to reduce the error of the predicted outcome of the NN model using forward propagation. Error or loss in an output is determined by traveling from output to back into the NN through backward propagation. This step is continued several times to minimize the error depending upon the error size. A complex round of forwarding and backward propagation is termed as an epoch. Stochastic gradient descent algorithms help optimization of error in NN model built for the huge dataset. Let us consider objective function as eq (6.2).

$$f(h) = \frac{1}{N}\sum_{i=1}^{N} f_i(h) \tag{6.2}$$

In eq (6.2), $f_i(h)$ is an error or loss function for a training set of N size, where N is too large for which computation cost can be too high. Stochastic gradient descent provides a solution for such huge dataset, that rather than using the whole gradient, it uses randomly created i number of small samples and calculate$\nabla f_i\,(h)$ as in the following equation:

$$\nabla f_i(h) = \frac{1}{N}\sum_{i=1}^{N} \nabla f_i h \tag{6.3}$$

To update h if λ is a learning rate where computation cost is $o(|\beta|)$:

$$h := x - \lambda\ \nabla f_i(h) \tag{6.4}$$

For large training samples, SGD is a powerful enough to find solution in few iteration. SGD minimizes the error in each iteration, where each of the iteration consists three steps:

Calculate the function closer to the output value as shown in the following equation:

$$f(h_{i)=} \sum_{i=0}^{N} \Omega_j f(h_j)\xi \tag{6.5}$$

Calculate error rate from the difference between actual value and $f(h)$ as shown in the following equation:

$$\Omega_i = y_i - f(h)_i \tag{6.6}$$

Error is minimized in each iteration by adjusting the coefficient of variable as shown in the following equation:

$$\Omega_i = \Omega_j + \delta\,(\in_i)h_{ij} \tag{6.7}$$

δ is a learning rate here and it should be very small for large dataset. Batch methods such as BFGS use entire data to calculate the next update which can be very slow and costly for a large dataset in the single machine, but stochastic gradient descent can overcome this challenge and provide fast convergence.

MLP model is constructed for prediction of frauds and nonfrauds of credit card transaction dataset. It is a large imbalance dataset consisting of approximately 3 billion records. Here MLP consists 1 input layer containing all the features of the dataset as node and bias, 1 hidden layer of 32 nodes, and 1 output layer with 2 nodes, each for 1 possible output. Before NN architecture is constructed, proposed algorithm is applied in a dataset for reducing its complexity. Then MLP is used on a balanced dataset to train the model. The NN is used here for building model to predict both frauds and nonfrauds accurately.

NN model training is started by initializing learning rate 0.01. RELU activation function is used as a unit of processing for the neural net as a type of activation function matters a lot for accurate training. Nowadays, most of the NN and deep-learning network are using RELU because it calculates the output in a simpler manner and it required less time to train large network.

Adam optimizer is used to calculate and then minimize the error.[13] Adaptive Moment Estimation optimization is a variation of stochastic gradient algorithm, requires less memory, and first-order gradient only. Adams calculate learning rate individually for each parameter from estimated moments of first and second gradients. We use five epochs during optimization.

Keras is a software used for construction of neural net architecture, for calculating output, for minimizing errors, and for evaluation of test prediction. Keras is a high-level software application written in python. Keras is composed of deep-learning library and API to be used. Tensorflow,

Theone, and CNTK act as a base for Keras. It gives a facility to convert an idea into result in very short time by focusing on fast experimentation. It is an efficient computational library for performing the numeric calculation. It helps to construct and train NN model with very less line of code.

6.12 EXPERIMENT II: HPRT-BASED NEURAL NETWORK MODEL

Credit card transaction dataset used during collaboration of ML group is considered here in this research. Dataset consists of 2-day credit card transactions of European card holder (2013). It is highly imbalanced dataset having 0.17% frauds only and 30 features. Much information about the dataset is not provided due to security reason. The intention here is to balance the dataset and then built ML model to catch the fraudulent so that increasing rate of frauds can be minimized by extracting necessary pattern used during the fraud.

MLP constructs a model for a balance credit card transaction dataset. Keras is used for training NN architecture. Nodes and layers are added in a Keras through the creation of the sequential model. We use three layers fully connected NN architecture. A number of layers can vary from problem to problem. Feedforward and back propagation technique is used for training purpose. Learning rate (0.1) is kept as learning rate because for high learning rate, time consumed for training increases while accuracy decreases. In a model, architecture number of nodes is the same as the number of features in a dataset. For the hidden layer, we increase the node by 2, for providing flexibility to a model but it can vary from analyst to analyst. In this study, we get a satisfactory result by adding one hidden layer with 32 nodes, which can again vary depending upon the problem and quality of its solution. Initially, we assigned weights to all the input nodes along with bias by using a random number generator. The training set is fed as input to the input layer of the NN. The output is passed with its weight from the input to hidden and then to the output layer. At the output, layer error is calculated by the Adam stochastic gradient descent algorithm and back propagated again into the input layer in case of error arises. The step is continued until the error is minimized. This step can be explicitly provided to the architecture by setting a number of epochs. Epoch is a number of iteration for which the training process will continue; here we use five epochs, that is, five cycles of feedforward and back propagation will happen to search and minimize the error thereafter reflects the final result. RELU is used as an activation function of input and hidden layer. Softmax calculates probability which ranges from 0 to 1, therefore, we use softmax as activation function of output layer because it produces result ranges only from 0 to 1, which is easy in mapping with target class 0 or class 1 depending upon the default threshold which is 0.5 for this experiment.

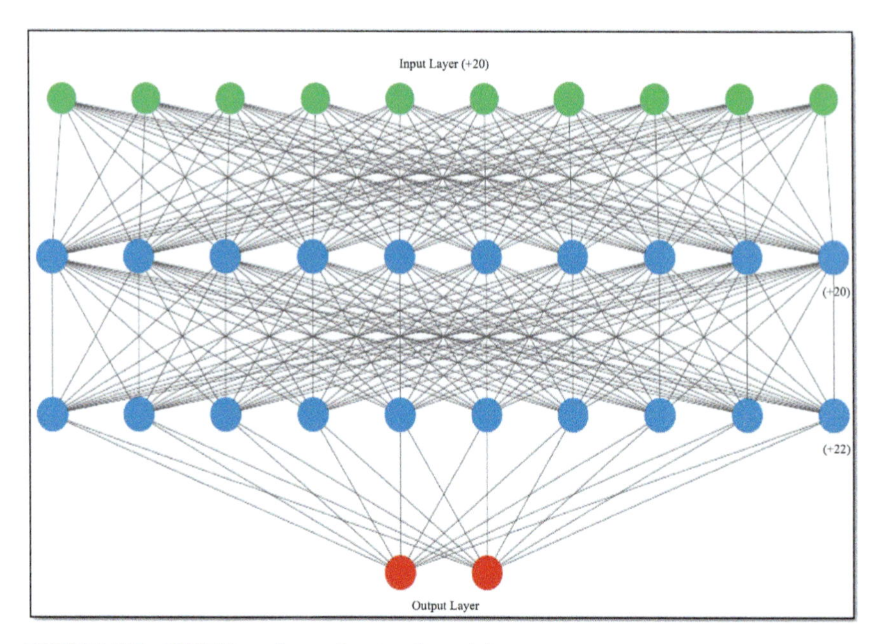

FIGURE 6.7 HPRT-based neural network model.

After processing of neurons, we must compile the model to calculate the loss. During compilation, we use sparse categorical cross-entropy for loss calculation and Adam optimizer for loss minimization. After compilation, we will check the model on the same dataset to determine the quality of the model. Here we will use epochs (no. of iteration) for improving the quality of our model by reducing errors. After training the NN model, the performance of the model is evaluated by testing it with test data. This will provide us an overview that how precisely our model captures the training rule. Further in future, if real-life unseen data is provided to the model it will predict output based on the training provided earlier.

```
Train on 243980 samples, validate on 60996 samples
Epoch 1/5
 - 4s - loss: 0.0192 - acc: 0.9928 - val_loss: 0.0036 - val_acc: 1.0000
Epoch 2/5
 - 3s - loss: 0.0036 - acc: 0.9989 - val_loss: 6.0133e-04 - val_acc: 1.0000
Epoch 3/5
 - 3s - loss: 0.0032 - acc: 0.9990 - val_loss: 0.0122 - val_acc: 0.9979
Epoch 4/5
 - 4s - loss: 0.0031 - acc: 0.9990 - val_loss: 0.0017 - val_acc: 0.9997
Epoch 5/5
 - 3s - loss: 0.0020 - acc: 0.9994 - val_loss: 6.8450e-04 - val_acc: 1.0000
```

FIGURE 6.8 Loss and error calculation during each epoch.

6.13 HPRT-BASED NEURAL NETWORK ALGORITHM

Input: HPRT applied dataset C

Learning rate: 0.01

Output: NN model

Method: Initialization of weights and biases for nodes of the input layer using random numbers.

While condition for termination not get satisfied

{

For each Feature T in C

{

Feed the training features to the input layer

Note: Feed forward propagation

For each node of the input layer I unit x propagation

{

Ix = Tx

For each node of hidden layer H at unit x

{

Hx = sum (WYxIx + αx)

Note: αx is a bias at H and WYx is a weighted input at layer Hx to previous layer y.

}

For each node of output layer O at unit x

{

Ox = Computed output from hidden layer Hx

}

For each node of output layer O at unit x

Ex= Ax – Ox

Note: Ax is an actual value from the training set

For each node from the output layer toward the input layer

{

Checking for the neuron causing a maximum error during back propagation

}

For each weight in network

{

ΔWyx = Adjust weight

Note: ΔWyx is a weight of higher connection to previous in network

WYx= ΔWyx

Note: weight update

}

For each bias in network

{

Δαyx = Adjust bias

Note: αyx is a bias of higher connection to previous in network

αyx = Δαyx

Note: Bias update

}

}

6.14 EXPERIMENT II: RESULT

MLP model prediction for fraud transactions has been compared with the test set sample, which is a part of the original dataset, extracted from it and kept unseen at the beginning of the experiment. It produces a very good result for various matrices and measures. Accuracy (99%), precision (100%), recall (85%), and *F*1 score (89%) results have been observed. According to confusion matrix out of 38,221 test samples, the model predicted, 38, 108-time TN, 87 times TP, 15 times FP, and only 11 times FN. Therefore, we can conclude the model is precisely predicting frauds reflecting too small misclassification and can be neglected.

TABLE 6.2 Results of Neural Network Models during Experiment II.

Matrices & Measures (%)			Traditional Neural Network Model		HPRT Based Neural Network Model	
Accuracy			99%		99%	
Precision			75%		100%	
Recall			78%		85%	
F1 Score			39%		89%	
Confusion Metrics						
Total Text Samples 38221	Predicted Negative	Predicted Positive	38112	11	38108	15
Actual Nagative	TN	FP	60	38	11	87
Actual Positive	FN	TP				

6.15 CONCLUSION

The NN works well when the data provided to it is balanced containing target class of normal distribution. But for an imbalance, dataset NN result is quite confusing, although accuracy (99%) is observed confusion matrix, exhibit lots of misclassification with very low *F*1 score (39%). Therefore,

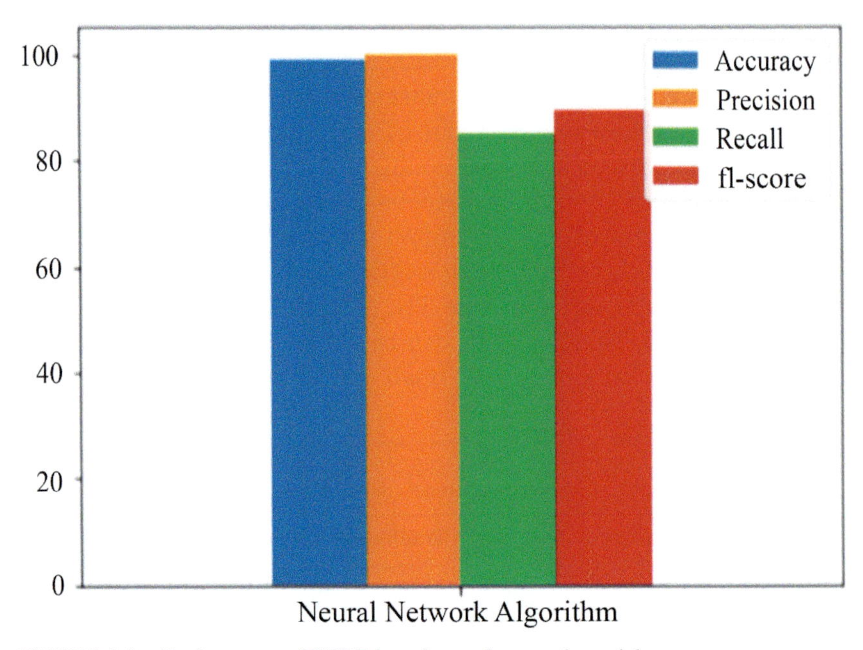

FIGURE 6.9 Performance of HPRT-based neural network model.

the traditional NN algorithm does not fit for an imbalanced dataset. In the second experiment, HPRT is applied on the dataset for reducing dataset complexity where the performances of NN algorithm also boost upbringing forth precise and accurate result with only 26 misclassifications out of 38,221 samples. HPRT acts as a preprocessing tool for a noisy and imbalance dataset, thus intensify the performance of ML classification algorithms. Finance industry could be benefited through this model by recognizing the fraudulent pattern and can alert in future to minimize losses happen through fraud transactions. This effort can save millions of dollars every year from bustling to the wrong person during fraud transactions. HPRT can utilize any imbalance dataset having a lot of redundancy and outliers but the algorithm can be extended further for removing other noises from the dataset so that the algorithm turns into more generalized and fit for any dataset. Enhanced HPRT-based NN model can utilize for prediction of credit card frauds for future data provided the dimensions of the dataset is same as that of the dataset used here.

KEYWORDS

- **machine learning**
- **data mining**
- **big data**
- **neural network**
- **data imbalance**

REFERENCES

1. Yadav, B. S. M.; Velagaleti, S. B. Challenges in Handling Imbalanced Big Data: A Survey. *Int. J. Curr. Eng. Sci. Res.* **2018,** *5* (3).
2. Zheng, Z.; et al. Oversampling Method for Imbalanced Classification. *Comput. Info.* **2015,** *34*.
3. Fitzgerald, J.; Ryan, C. A Hybrid Approach to the Problem of Class Imbalance. *Int. Conf. Soft Comput.* **2013,** *12* (1).
4. Fan, J.; et al. *Principal Component Analysis for Big Data*, January 2018, Accessed: https://arxiv.org/pdf/1801.01602.pdf.
5. Jain, M.; Verma, C. Adapting *k*-Means for Clustering in Big Data. *Int. J. Computer Appl.* **2014,** *101*.
6. Xi, J. *Outlier Detection Algorithm in Data Mining*. In 2008 Second International Symposium on Intelligent Information Technology Application, December 2008.
7. Vijay, P.; Keshwani, B. Emergence of Big Data with Hadoop: A Review, IOSR. *J. Eng.* (IOSRJEN), **2016,** *6*(3).
8. Radha, R.; Murlidhara, S. Removal of Redundant and Irrelevant Data From Training Datasets Using Speedy Feature Selection Method. *Int. J. Comput. Sci. Mobile Comput.* **2016,** *5* (7).
9. Sharma, R.; Gulati, N. Improving the Accuracy and Reducing the Redundancy in Data Mining. *Int. J. Eng. Sci. Comput.* **2016,** *6*.
10. McCulloch, W. S.; Pitts, W. A Logical Calculus of the Ideas Immanent in Nervous Activity. *Bull. Math. Biophys.* **1943,** *5* (4).
11. Mehta, A. J.; et al. A Multi-Layer Artificial Neural Network Architecture Design for Load Forecasting in Power Systems. *World Acad. Sci., Eng. Technol. Int. J. Electr. Comput. Eng.* **2011,** *5* (2).
12. Bottou, L. *Large-Scale Machine Learning with Stochastic Gradient Descent*. NEC Labs America: Princeton, December 2016; Vol 5 (4).
13. Kingma, D. P.; Ba, J. L. *Adam: A Method for Stochastic Optimization*. Published as a Conference Paper at ICLR, Vol. 5 (2), June 2015.

14. Vijay, P.; Keswani, B. Support Vector Machine (SVM) Kernels Based Approach for Detection of Breast Cancer. *CASS* **2018,** *2* (2).
15. Vijay, P.; Keswani, B. A Study on Big Data Analytics through R. *Int. J. Innovative Res. Comput. Commun. Eng.* **2016,** *4* (8).

CHAPTER 7

Artificial Intelligence in Education Using Gaming and Automatization with Courses and Outcomes Mapping

S. MANIKANDAN* and M. CHINNADURAI

E.G.S. Pillay Engineering College, Nagapattinam, Tamil Nadu, India

Corresponding author. E-mail: profmaninvp@gmail.com

ABSTRACT

Nowadays, education is important for all humans, and a variety of online-based solutions are developed for improving education and learning by computer. Artificial intelligence (AI) is an important development and an automated tool for motivating humans or learners in education. There are many AI-based games and applications that are being developed like competition, ranging, real-time applications, educational applications, natural language processing, and so on. All of this is possible because of the various materials and open-based mobile applications that are developed for different human activities. This chapter discusses the study of the involvement of AI in educational field, and various theories are analyzed, automated by multiagent systems. We developed an empirical study of different gaming models and automation systems used for the processing of parallel applications. The autonomous nature of AI is implemented for different commercial applications and apps like Angry Birds and Age of Empires for gaming, social network apps used for social interactions, and educational apps that are used by the learning community. With the technological evolvement and development comes a competition of different automated environments. From this competition, we consider the following outcomes, such as computer vision applications, decision-making, strategic planning, resource allocation, and management applications. The comparison of various real-time applications is analyzed with different outcomes, and this provides state-of-the-art gaming theory and is applied to provide an exact survey of AI in education. We promote the understanding of AI in educational society in comparison to human approach, and based upon the same, we provide autonomous solutions. Furthermore, we project that AI in education will be more competent, and various optimizing procedures will make social applications more reliable.

7.1 INTRODUCTION

Games are one of the major agents or applications to motivate the student community and provide good platforms for learning and education. Most of industries have developed various mobile apps games for learning perception and improve the IQ level of the students. According to the survey of Intelligent Agent Conference in 2012 held at the USA, 90% of the students use different mobile apps; as a part of the revolution, the top companies, such as Microsoft, IBM, Apple, and so on, are designing automated and autonomous natured games educating the students and competing with each other by upgrading them constantly.[1] As per the survey report by Kim et al., Playstore is the first platform to provide Android-based mobile apps in which 67% are occupied gaming apps and 45% of gaming applications are used in education.[2] The various gaming applications are classified by thinking nature, playing nature, adventure nature, and video-related commands or instruction level applications. Different kinds of levels are used for thinking and the ability to act intelligently.

Games are developed of physical nature and properties of each playing object have unique and certain behaviors.[3] Recently, the intelligent industry conferences included the following characteristics that are important for developing learning applications: (1) competition-based games are used to thinking different levels; (2) racing and adventure-based games are used to improve the brain accuracy levels; (3) the hand or finger-inputting board games are used to understand the capabilities. So the organizers rely much on interviews to provide software as a service platform that is used to build AI-based education games. For the development, the participants can use rule-based approach or artificial-based platforms are used with predefined characteristics.[4]

With the demand for intelligent and automated-based applications, the following psychological behaviors are needed to check learning abilities like cognitive basics of behaviors, instance of memory, learning aspects, evolution, and forecasting events. Information metabolism theories are used to describe human emotions after the same is played or while playing games. In this chapter, the behavior of humans is investigated and mental components, constraints, actions, and dynamic changes are parallelly analyzed. The intelligent neural network has been used to analyze the hybrid nature of performing subtasks, replacement of existing characteristics, and presented interest among the user.[5] In addition, the basic developments of software tools to deploy the competition are explained at each level.

The goal of the competition is used to build and entail analyzing the studies and building characteristics of various application models. In 2015, analyzing various surveys following behaviors are listed, such as image processing, resource allocation and management, and simulations parameters and results.[6] In this chapter, clear details about artificial intelligence in the education field with courses-and-outcomes results are given. In this chapter, other sections are described as follows: Section 7.2 explains the background details of artificial intelligence (AI) in education; Section 7.3 gives information about processing and gaming theory; and lastly, section 7.4 describes experimental setup and simulation results with conclusion.

7.1.1 BACKGROUND

The cognitive natures of abilities, such as various inference, messaging, learning procedures, and intelligent behaviors solve the new aspects of learning skills acquiring different tasks. AI systems have various aspects, for example, semantic nets with decision-making and perception. Analysis of the various components of nature set of axioms is listed.

1. The semantic, cognitive, mathematical, and algebraic structures are desired to implement AI models.
2. Database is used as a long-time memory which helps to gain knowledge and produce various functional results.
3. The rule-based consequences can be expressed by inference rules.
4. Set of modeling inputs, solving specific problems, and operation handlers are needed to make effective decisions.

Task-solving and teach pendant agents are first implemented with learning capabilities and used in the education field. This structure can model time-based learning practice and artificial capabilities.[7]

The various information processing system in autonomous behaviors, but AI agents have cognitive performance and information-handling capabilities that increase the level of services.[8]

In this chapter, we calculated and adopted some basic procedures and principles. The AI components are used in commercial and open-source platforms. UI-based behaviors provide common platforms for handling coding and implementation. There are common platforms designed to play the games and apply trajectory-based courses and outcomes. The following are the cognitive modules for knowledge processing (Fig. 7.1).

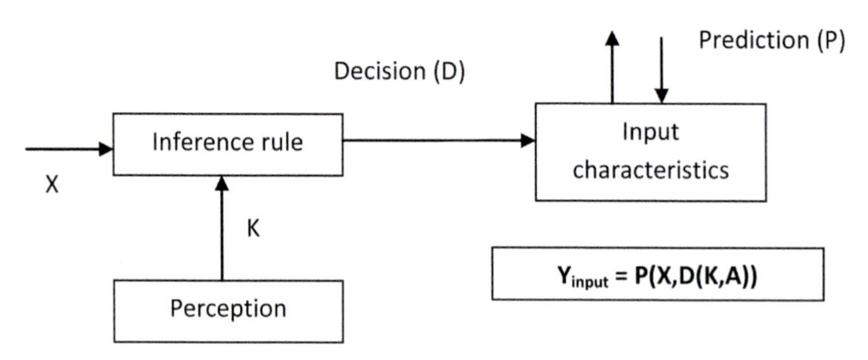

FIGURE 7.1 Knowledge representation formula.

For each decision-making process, a large number of input systems are applied for calculating the following simulation results:

1. To perform congestion operation to hold the weak inference relation behaviors
2. To set various problem statements as input and to calculate reflection mechanisms
3. To replace step-by-step automated function to calculate nearby nodes

A number of studies can be applied for single and multiple complex behaviors to calculate real-time AI constraints.[9] This chapter describes the empirical analysis of AI in education that can be experienced by various systems.

7.2 INFORMATION THEORY AND AUTOMATIZATION

In this chapter, we describe gaming information processing and automatization procedures that are applied for calculating education applications. Hierarchical is the process of step-by-step incremental procedure to solve or apply learning procedures with the following levels such as logical reasoning, natural-language processing, emotional intelligence, and unconditional characteristics. The learning levels can be achieved by set of automatic reactions, continuous levels of sources, and set of psychographic effectiveness.[10] Most of the consumption and real-time order-processing system will have the following features:

1. Automatic reaction and sustained optimized value processing are needed.
2. The structure of information process model is designed by stability and dependent functional behaviors
3. Mathematical procedures are needed to set cognitive rules and inference behaviors
4. This process is involved to set automation and set of time and energetic procedures

So the automatization process time varies and information are processed with physical fields, learning capabilities, and role-based conditions.

The following are the impulsive process model with inference behaviors (Fig. 7.2).

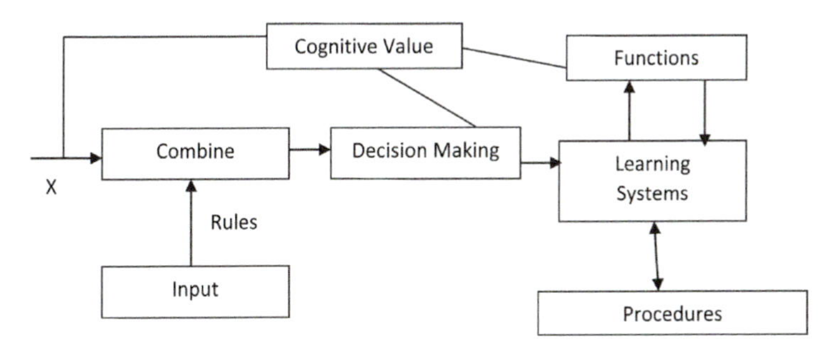

FIGURE 7.2 Cognitive value processes with inference rule and decision-making.

According to information processing, system will have a set of actions and ways of organized learning system. The following results can be gained from the set of information processing operations:

1. Collecting perception information from sensor as input and apply inference rule
2. Combining the rules and apply cognitive procedures and train the motivation process or collection information process
3. Appling decision-making capabilities with learning system and functional characteristics
4. The recognized information are noticed by applying different levels of hierarchical characteristics
5. Linking and learning process applicable for each bad input conditions

Various emotional levels in learning process can be analyzed and natural learning process can be applied by using the information in Figure 7.3.

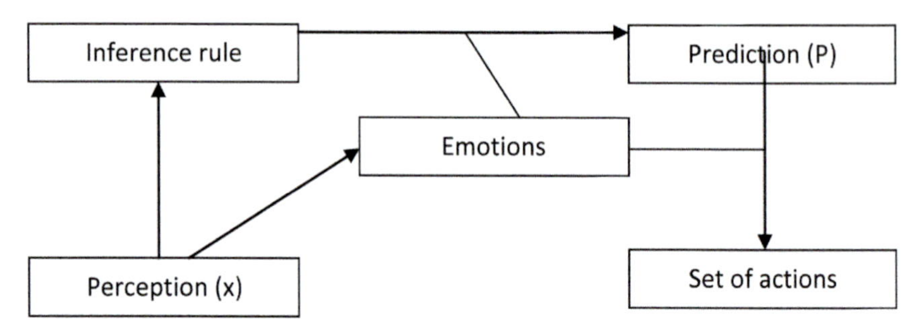

FIGURE 7.3 Emotion processing systems.

In Figure 7.3, let us consider the set of X is the input values and it enables various control information. The following are the automatic process with information algorithms:

Let K—Set of cognitive values and P—inference algorithm characteristics

A & B—set of possible input and output optimized results

G(V,P) = {A,B}

Set of conditional probabilities reactive $R^A \rightarrow R^B$ is used for analyzing behaviors. The following are the vision-based measures of computing learning systems such as object-oriented programming, operating systems, and so on. Table 7.1 shows the course and outcomes summary report.

The human can improve various performances by agent-based learning process. For the output of reflective modules calculated by (B_{out}, A_{inp}) € $R^A \rightarrow R^B$. The learning process is stopped if the output modules are equal to cognitive and control beginnings and actively participates with respect to time constraints and conditions.

TABLE 7.1 Courses and Outcomes Summary Report.

Course	Summary	Tool
Computer technology	Gaming and decision-making	Intelligent systems
Object-oriented labs	Gaming theory	Framework engine
Simulation	Interesting and puzzles	SIMulate
Operating systems	Learning capabilities	Peer interactions
Java programming	Web based and multiuser	Web or mobile apps engine
AI and deep learning	Learning system	Machine learning

7.3 IMPLEMENTATION OF AUTOMATIZATION AND GAMING THEORY

In this section, the result can be applied based on automated and gaming model with learning process, and it takes the input perception from the following impacts such as visual analysis, unnecessary behaviors, and so on. Table 7.2 shows that the activities chart for courses and outcomes mapping.

TABLE 7.2 Activities and Outcomes Chart.

Activity	Outcomes
Lecture	Introduction
Human actions	Step-by-step values
Benchmarks	Levels and attainment
Completions	Organized or arranged
Reviews	Analysis
Assessment	Test and decision
Feedback	Improvement
Evaluation	Performance
Questionnaires	Qualification

The reactions cannot be processed or left by the user because the automated process is handled in each stage with recorded values. The following process is used to measure automated processes and produce the monitored results (Fig. 7.4).

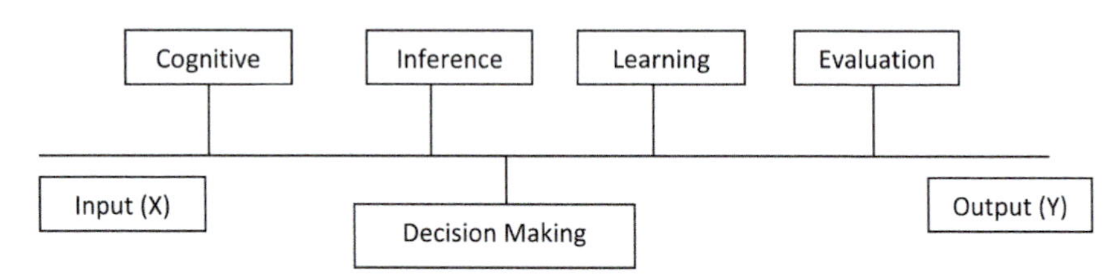

FIGURE 7.4 Automatization decision-making process.

A standard rule-based consolidated database with sample code is processed and handled with multiple requests at the same time. The following constraints are set and allow calculating cognitive values:

- Frequent use of demand process
- Real-time order processing or request handlers
- Rule-based learning processor or multiagent inference systems

The following are the customized level of each handling process and machine learning behaviors:

1. The maximum utilization of time and CPU overhead can be calculated and verified at each stage
2. The distance can be calculated by fuzzy constraints and input commands
3. The nearest next node values can be updated at any time and which is used for further order processing
4. An empty node, active node, processing node, and resultant node can be calculated scene behaviors

The performance of the AI agent can be analyzed by a set of questionnaires based on feedbacks and analysis (Table 7.3).

TABLE 7.3 Set of Sample Questionnaires for Feedback and Reviews.

The following are the set of questions that are set by mapping courses and outcomes:
1. What is the use of courses and outcomes in AI?
2. How to play the game and provide the solution for competing with the computer?
3. Are you satisfied with UI-based inputs on I/O?
4. What learning content is available at each level?
5. List the calculation parameters in each node's values?
6. What are the important problems or issues that arise while playing?
7. Differentiate satisfactory and unsatisfactory parts?

Based on this set of questionnaires, we analyzed various users' inputs and we made effective decisions for processing the next level. The negative-based answers are considered more important for calculating attainment and random auditing is done for the verification and validation process. Team-based answers are carried out for further reading.

7.4 CONCLUSIONS

The designs of AI-based education application are very useful for students and learner community to master solving complex problems and improve their IQ levels. We proposed courses-and-outcomes-based comparison survey results with gaming and information-processing systems. The use of AI in the real market and its competition significantly reduce the time and resource process, and automatization process is useful for defining problems, benchmarks, questionnaires, and feedback-analysis stages. This chapter includes a description of decision-making capabilities, computer-vision-based processing, ordering and delivery reports, expert system-based mobile applications, and how to handle uncertain situations while playing or working. The result counter files relationship between strong and weak agent's behaviors at certain and uncertain conditions. It shows that learning and motivation are important factors for running AI applications or expert systems with intelligent decision-making behaviors. All the users' feedbacks are taken into account for the improvement of the next level and it will continue further for approaches in different reasoning-based techniques.

KEYWORDS

- **artificial intelligence**
- **courses and outcomes**
- **gaming theory**
- **autonomous system**
- **decision-making**

REFERENCES

1. Yoon, D.-M.; Kim, K.-J. Challenges and Opportunities in Game Artificial Intelligence Education Using Angry Birds. *IEEE Transl. Content Min.* **2015,** *3*. DOI:10.1109/access.2015.2442680. ISSN: 2169-3536.
2. Bielecki, A. A Model of Human Activity Automatization as a Basis of Artificial Intelligence Systems. *IEEE Trans. Autonomous Ment. Dev.* **2014,** *6* (3). ISSN: 1943-0604.
3. Kudo, F.; Akitomi, T.; Moriwaki, N. An Artificial Intelligence Computer System for Analysis of Social Infrastructure Data. In 2015 IEEE 17th Conference on Business Information; pp 85–89. DOI: 978-1-4673-7340-1/15.
4. Deng, L. Artificial Intelligence in the Rising Wave of Deeping Learning the Historical Path and Future Outlook. *IEEE Signal Process. Mag.* **2018.** ISSN: 1053-5888@2018.
5. Tang, Y.; Xing, X.; Karimi, H. R.; Kocarev, L.; Kurths, J. Tracking Control of Networked Multi-Agent Systems under New Characterizations of Impulses and Its Applications in Robotics Systems. *IEEE Trans. Ind. Electronics* **2016,** *63* (2), 1299–1307. DOI:10.1109/tie.2015.
6. Su, S.; Lin, Z. Distributed Consensus Control of Multi-Agent Systems with Higher Order Agent Dynamics and Dynamically Changing Directed Interaction Topologies. *IEEE Trans. Automat. Control* **2016,** *61* (2), 515–519. DOI:10.1109/tac.2015.2444211.
7. Hu, J.; Feng, G. Distributed Tracking Control of Leader-Follower Multi-Agent Systems under Noisy Measurement. *Automatica* **2010,** *46* (8), 1382–1387.
8. Carli, R.; Chiuso, A.; Schenato, L.; Zampieri, S. Optimal Synchronization for Networks of Noisy Double Integrators. *IEEE Trans. Automat. Control* **2011,** *56* (5), 1146–1152.
9. Ma, C.; Zhang, J. Necessary and Sufficient Conditions for Consensus Stability of Linear Multi-Agent Systems. *IEEE Trans. Autom. Control* **2010,** *55* (5), 1263–1268.
10. Fountain, J.; Walker, J.; Budden, D.; Mendes, A.; Chalup, S. Motivated Reinforcement Learning for Improved Head Actuation of Humanoid Robots. In *RoboCup2013: Robot Soccer World Cup XVII*; Springer: Berlin, Germany, 2013.

CHAPTER 8

The Impact of Artificial Intelligence for Comprehensive Growth Across the Globe

SHALINI SRIVASTAV[1], VIKAS GARG[2], and AHMED A. ELNGAR[3,*]

[1]*Commerce and Finance, Amity University Uttar Pradesh India, Greater Noida Campus, India*

[2]*Director, Executive Programs Management, Amity University Uttar Pradesh India, Greater Noida Campus, India*

[3]*Faculty of Computers and Artificial Intelligence, Beni-Suef University, Beni-Suef City, Salah salem str., 62511, Egypt*

[*]*Corresponding author. E-mail: elngar_7@yahoo.co.uk*

ABSTRACT

It is a great decline in the ability to propel the economic progress through increased capital investment and labor. These two levers are classics. These two levers are the traditional production drivers, yet they cannot sustain a steady growth of positivity which is enjoyed in most developed economies in previous decades. The "new spring" for artificial intelligence (AI) has prompted much work into AI's economic effects, and there is an emerging consensus that it will deliver significant benefits. Research tells us that a wide range of technologies can increase the profitability and very high levels of GDP growth. The same data numbers vary as researchers have used different methodologies, for example, analysis of an economic effect small or wide range of drivers.

A large variety of businesses are now using artificial intelligence software in a broad range of forms and functions. From early 2018, supply chain industry and fixed assets like developing the preventive maintenance of was taken care of by AI that use artificial auditory cortexes to reduce the perception of human speech, and can therefore automate the recognition and analysis of sources of possible system breakdowns, for example, the use of AI by Quantum Black to streamline R&D in Formula 1 racing ; and sales and marketing too. Yet, pessimism in the long run is unwarranted.

The transformative set of technologies as per the recent changes in which artificial intelligence is having the potential to find out the restrictions physically in case of money and labor and to find out new areas of growth and ethics in order to enter a new phase.

8.1 INTRODUCTION

In search of a better solution, policymakers and corporate leaders need to plan and move for a future with artificial wisdom. We will not do so for the concept that artificial intelligence (AI) is another medium to enhance the efficiency of the system. Moreover, AI is considered as a tool that can change our way of thinking and increase creativity. Gross domestic product (GDP) growth rates have been declining across the globe. That was valid for three decades. Key economic productivity indices show a sharp downward trend, while labor-force growth in the developing world is essentially sluggish. In some countries, pundits are already in recession suggesting despite this weak outlook, that the "new" economy is a static economy.[1]

The total GDP in 2016 is estimated to have hit around $75 trillion. Our baseline estimates indicate that by 2030, the number is valued at about $114 trillion. Our study of the S-CGE model indicates that the global GDP in 2030 will reach 14% higher as a result of the use of AI which is equivalent to 15.7 trillion.[2]

Economist Robert Gordon argues, on an even more pessimistic note, that the productivity growth will be very slow for the next quarter century as we all are seeing way back year 2004. He believes that the last two centuries of "Great Inventions," such as steamships and telegraphs are unlikely to repeat themselves. And combined with adverse population patterns, this productivity gap is flagging educational achievement, and rising income gaps would delay economic growth. That is how the economic growth is impacted by new technologies. Money and labor are traditionally the "production factors" which drive the economic growth. Development takes place as capital supplies or labors increase, or where they are used more efficiently. In total productivity factor (TFP) is caught the development that comes from inventions and economic technical transition.[3]

It is seen that economists always consider new technologies as drivers in the growth process by enhancing TFP. That made sense with the technologies we have been witnessing so far. In the last century, major technical breakthroughs, power, rail, and IT improved the efficiency of technology. We suggest that a variation of this automated growth feature can yield a rich description of the growth process and this has the implications for economic development and the allocation of wealth. The application of a model AI automates the production of products and services, an idea from Baumol generates enough conditions under which the overall sustainable growth can be accomplished with a constant share of capital that remains just below 100%, even with almost full automation.[4]

Today, we are seeing another ground breaking series of take offs, generally referred to as AI. Many people consider AI as proximity to past innovations. If we think that this is something, we should expect to expand but nothing revolutionary. The goal is to find out AI as a labor–capital hybrid. AI can simulate labor operations even higher in size and time, and even execute tasks beyond human ability. Like robots and smart machines, AI can be similarly used in the form of a physical capital and in contrast to traditional property, such as machines and houses, and due to its ability to self-learn, it can change over time. Based on our research and simulations, we are able to demonstrate what happens when AI is used as a major element in development rather than a pure efficiency enhancer. For example, the impact on projected growth for the US is dramatic.

In just 2 years, AI is tipped to raise India's rate of innovation by around 230%, based on estimates. In recent times, the country has been concentrating on research but is still far from keeping up with the world leaders.

A national AI strategy must be based on a structure that is tailored to the specific requirements and expectations of the country, according to NITI Aayog's discussion paper on AI's National Strategy for AI. At the same time, it is capable of achieving the maximum potential of exploiting AI technologies in India.

In an attempt to make AI a reality for India, a proposal to establish an institutional structure was proposed by NITI Aayog in May this year. A cabinet note was circulated to provide Rs 7500 crore in funding for the development of a cloud computing platform called AIRAWAT and research institutes, initially over a 3-year period, and to set up a high-level task force to oversee the country's rollout and implementation of AI.

AI has the capacity to accelerate development from an economic impact perspective by automating multilayered physical activities requiring adaptability and agility across industries. It also enables the human workforce to concentrate on their position, which can add the most value, complement human capacity, and can increase the capital productivity.

Accenture offers a framework for evaluating the economic effect of AI for select G20 countries in its latest AI research report and forecasts that the technology can boost India's annual growth rate by 1.3% points by 2035. The technology has the potential to add US$ 957 billion, an improvement of 15% of the current gross value of India in the same year.[5]

In its national AI strategy paper, NITI Aayog has also assessed a large number of sectors where AI can be successful and has taken a deliberate decision to concentrate on a small collection of sectors where only initiatives led by the private sector would not lead to the desired social outcomes.

The target sectors include (1) education-preparing for the next generation to take advantage of India's global AI revolution, (2) smart cities and infrastructure development for the rapidly urbanizing population of India, and (3) smart mobility and transport-solving problems, such as congestion, pollution, high road accident rates leading to economic inefficiency, and massive human costs.

This paper looks at the economic development effects of AI. AI may be defined as "a machine's ability to imitate intelligent human behavior" or "the ability of an agent to achieve objectives in a wide range of environments." These meanings instantly invoke fundamental economic problems. What happens, for instance, if AI will make it possible to automate an ever growing number of activities traditionally carried out by human labor? In the ordinary production of goods and services, AI can be deployed, thus potentially affecting the economic growth and revenue shares. But the process by which we create fresh ideas and technologies using AI may also change, thus helping us to solve complex problems and scaling up creative efforts. Some observers have argued that in extreme versions, AI can rapidly become self-improving, leading to "singularities" featuring unlimited machine intelligence and/or finite-time unlimited economic growth (Good, 1965; Vinge, 1993; Kurzweil, 2005). Nordhaus (2015) provides a detailed overview and discussion of the prospects for uniqueness from the perspective of economics.

S.No	Topic	Name	Journal name	Literature review
1	Assessing the Economic Impact of Artificial Intelligence	Jacques Bughin, MGI Director and Senior Partner of McKinsey & Company, Jeongmin Seong, Senior fellow, MGI, and MGI's expert members	ITU Trends: Emerging trends in ICT'S, Issue Paper No.1, September 2018	Front-runners are described as businesses that over the next 5–7 years implement a wide range of AI technologies and absorb the application of technologies through their organizations. The simulation presumed that this category consists of approximately 10% of companies whose AI-facilitated growth profile is comparable with that of the top quartile of high-growth performing companies. This segment is similar in spirit and scale to the early adopters found by Everett M. Rogers, among others, in the theory of technology diffusion
2	The Past Decade and Future of AI's Impact on Society	Martin Rees , Francisco Gonzalez and team	Towards a New Enlightenment? A Transcendent Decade- Book 2019	Artificial intelligence (AI) is a term used to describe objects that respond to observed contexts in order to identify contexts or influence behavior. Our ability to build such devices has progressed, as has their impact on society. This article first examines the social and economic developments brought about by our usage of AI, with an emphasis on the decade following the introduction of smartphones in 2007, which contributed significantly to "big data" and hence to the effectiveness of machine learning. It then makes predictions, including policy recommendations, about the political, economic, and personal issues that humanity will face in the near future
3	Why artificial intelligence is the future of growth?	Mark Purdy and Paul Daugherty	Why Artificial Intelligence is the Future of Growth— Accenture Canada	Unlike conventional automation solutions, the latest AI-powered wave of intelligent automation is already generating growth across a collection of characteristics. Its ability to automate the complex physical world activities requiring adaptability and agility is the first feature. Consider the job of retrieving goods in a warehouse, where businesses have depended on the capacity of people to negotiate cramped spaces and prevent obstacles from moving. Today, lasers and 3D depth sensors are used by Fetch Robotics robots to maneuver safely and operate alongside warehouse staff. Used in tandem with humans, the robots in a typical warehouse can handle the vast majority of products

S.No	Topic	Name	Journal name	Literature review
4	AI will create as many jobs as it displaces—report		BBC News, 16 July 2018	According to PwC's estimate, AI will promote economic development by creating new jobs while old ones fade away. However, it warned that the industry sector will have "winners and losers," with many positions likely to change. Opinions on AI's potential influence are mixed, with some predicting that technology will put many people out of work in the future. Pessimists claim that AI is unlike any other technological channel before it
5	Realizing the growth potential of AI	Stephen Diorio and thought leaders	Forbes	AI and Machine Learning (ML) will alter enterprises by lowering costs, minimizing risks, streamlining processes, speeding growth, and sparking innovation according to business leaders and investors. The potential for AI to increase sales and profit is huge. According to a Forbes survey of 1093 executives, marketing, customer service, and sales are the top three functions where AI can reach its full potential
6	How the future of work changed in 2020		A Pega Report	What's driving digital change? And what instruments are businesses using to help them adapt? To find out, we surveyed more than 3000 senior business leaders about the emerging role of technology in the workplace
7	Bridging the perceived gap between industry and academia	Dr. Shalini Srivastav, Dr. Vikas Garg, Dr. Anubhuti Gupta	International Journal of Supply Chain and Operations Resiliance, 2020 Vol.4. No.2	Need is creation's mother. So is the cooperative initiative of the academy—industry, handled by need. Over the last decade and a half, the planet has turned into a global city. Employers today operate in a domain that calls for new and continuously evolving skills to sustain global intensity. Nevertheless, regardless of the awareness of the value of such partnerships across the board, ironically, such collaborative efforts are very limited in India as well as all over the world. Due to various barriers to concerted efforts by the academic sector that still exist, the reason behind this can be attributed to the absence of a separate model

S.No	Topic	Name	Journal name	Literature review
8	How artificial intelligence is transforming the world	Darrell M. West and John R. Allen	Brookings Report	AI is a broad technology that enables people to rethink how they incorporate data, interpret data, and apply the ensuing insights to improve decision-making, and it is already transforming every aspect of life. In this article, Darrell West and John Allen look at how AI is being used in a variety of industries, how to fix problems with its expansion, and how to get the most out of AI while maintaining important human values
9	Seven ways AI can change the world for better … or worse	Mike Thomas	The Future of Artificial Intelligence, April 20, 2020	IFM is just one of several AI pioneers in an area that is hotter than ever and becoming hotter all the time. Here is an example of an excellent indicator: AI-related inventions accounted for 1.600 (or over 18%) of IBM inventors' 9100 patents in 2018. Here is a new one for you: Elon Musk, the founder of Tesla and a tech tycoon, has gave $10 million to the OpenAI nonprofit research business, a drop in the bucket compared to his $1 billion co-investment pledge in 2015. "Whoever becomes the top in this field [AI] will become the world's ruler," Russian President Vladimir Putin warned schoolchildren in 2017. He then threw his head back and burst out laughing
10	The impact of artificial intelligence on employment	Georgios Petropoulos	The Impact of Artificial Intelligence on Employment	It has become an integral part of the most demanding and fast-paced markets, less than 70 years from the day the very word AI came into being. In order to get a competitive advantage on the industry, forward-thinking executive managers and company owners aggressively pursue new AI usage in finance and other fields
11	On the effects of artificial intelligence on growth and employment	Philiphe Aghion et al.	Economy – Global Economy	If combined with an inappropriate competition policy, AI can inhibit development. The effect of robotization on jobs in France over the period 1994–2014 is addressed in the second part of the paper. Based on our empirical study of French results, we demonstrate, first, that robotization decreases aggregate jobs at the level of the employment zone, and, second, that robotization affects noneducated workers more negatively than educated workers. This result indicates that the positive effect that AI and automation could have on jobs is limited by inadequate labor market and education policies

S.No	Topic	Name	Journal name	Literature review
12	Global economy implications of artificial intelligence	Priya Dalani, February 2020	Analytics Insight	A more influential number of individuals, companies, and governments are adopting AI and ML as growing performance and productivity enables exponential growth in specific areas of the global economy. However, the productivity and efficiency disparity between AI and ML industries and organizations versus those that have not increased exponentially. With less and less hope of making up for lost time with the pioneers, this risks leaving us at the foundation far behind.
13	The economics of artificial intelligence	Philippe Aghion, Benjamin F. Jones, and Charles I. Jones	National Bureau of Economic Research	Another subject in our chapter is that Baumol's "cost disease" which may limit the growth effects of automation and AI. According to Baumol (1967), sectors with rapid productivity growth, such as agriculture and even industry today, frequently see their share of GDP drop, whereas sectors with relatively modest productivity growth may see their share of GDP increase gross domestic product (GDP). As a result, economic progress may be constrained by what is necessary but difficult to alter, rather than what we are doing effectively.
14	Economic impact of artificial intelligence: New look for the macroeconomic assessment in Asia-Pacific region	Muhammad Haseeb, Sasmoko, Leonardus W. W. Mihardjo, Abid Rashid Gill, Kittisak Jermsittiparsert	International Journal of Computational Intelligence Systems	A second theme in our chapter is that the "cost disease" of Baumol may restrict the growth consequences of automation and AI. Baumol (1967) noted that sectors with rapid productivity growth, such as agriculture and even manufacturing today, often see their share of GDP decline, while those sectors with relatively slow productivity growth may include their share of GDP. As a result, economic growth could be limited not by what we are doing well, but rather by what is necessary and yet difficult to change.
15	The role of artificial intelligence in achieving the sustainable development goals	Ricardo Vinuesa,	Nature Communications volume 11, Article number: 233 (2020)	Techniques, such as statistical learning, ML, object detection, processing of natural language (NLP), computer vision, speech recognition, and pattern recognition have been employed by AI. In different activities, such as sensing vehicle distances, recognizing speech in audio recordings, recognizing faces in images, and translating languages, these techniques are used. AI applications help in inducing some degree of "intelligence" to give computers space to make sense of the data they are presented with

S.No	Topic	Name	Journal name	Literature review
16	Toward understanding the impact of artificial intelligence on labor	Morgan R. Frank et al.	Proceedings of National Academy of Sciences	There is the potential for rapid developments in AI and automation technology to dramatically disrupt labor markets. While AI and automation will improve the efficiency of some employees, they can reduce the work done by others, and at least to some extent, would likely change almost all occupations. In a time of increasing economic disparity, rising automation is growing concerns of mass technological unemployment and a renewed demand for policy efforts to tackle the effects of technological change. The obstacles that hinder scientists from assessing the impact of AI and automation on the future of work are explored in this paper
17	Artificial intelligence, income distribution and economic growth	Thomas Gries and Wim Naudé	IZA Institute of Labor economics	The net impact of automation on employment would therefore depend on the relative strengths of the displacement effect and the countervailing reinstatement effect. An underlying assumption is that AI-driven innovation would be distinct from other forms of automation (e.g., robots) in being more likely to produce countervailing reinstatement effects. Nevertheless, Agrawal et al. (2019) is different from this assumption. They clarify that AI is basically a prediction technology and that it will mainly displace labor from prediction-required tasks and that it will only create (re-establish) jobs elsewhere in tasks that can be done better with the help of superior prediction technology and tasks that can be done better that will benefit upstream or downstream from these new tasks
18	Artificial intelligence for inclusive growth	Amit Kapoor and Harshula Sinha	The Economic Times—E Paper	AI adoption is not limited to companies, but economies have also shifted their attention to building up their AI capabilities as a means of growing growth. Developed economies are already in the race, such as the US, China, and EU countries. India is now set to join them, too. The Government of India has formed a Task Force on AI over the last few years and mandated the NITI Aayog to prepare a National AI Strategy to exploit AI for inclusive development

S.No	Topic	Name	Journal name	Literature review
19	How can AI help in reviving the economy?	Shraddha Goled	Analytics India Magazine	AI-enabled personalization: Personalization of goods and services can be significantly enhanced by the use of AI. Automating routine tasks by robots and autonomous vehicles would boost productivity gains, as per a PwC report. Eventually, the availability of customized goods and services would also increase demand, which will produce more data in turn. Increased supported, autonomous, and augmented intelligence for investment would help the workforce reduce the pressure on routine tasks and focus more on higher value-added activities instead
20	Role of artificial intelligence in Indian economy	Role of artificial intelligence in Indian economy	Centre for Economic policy and Research	As the entire globe is entering a new technology phase, it is necessary for our country to keep innovating and upgrading its technology accordingly in each field. AI is one such area that contains a great deal of opportunity and good fortune if India is able to use all of its ability and at the same time has a bad fortune if it is unable to meet the level of technology demanded by the prevailing level of competition. It is necessary, therefore, to take corrective measures and initiate policies that this sector needs. Looking at the current problems facing this field, a great deal of government effort is required

8.2 DEVELOPED ECONOMICS: THE GROWTH ENDS

On various main areas, the economic data are supporting the long-term pessimistic approach. Since 1980s, the GDP has majorly reduced in many economic countries.

Capacity: A main system of how efficiently the countries are using their money and people in full capacity. The data below tell us about the slow-down of companies in the past 10 years.

Efficiency Capital: It is seen that the marginal rate of capital efficiency, which indicates the increase in capital profitability in areas, such as machines and buildings has drastically reduced in the last 50 years.

Labor: When demographics mature and birth rates fall, there are fewer topics left in the workplace up to the slack.

Machine learning (ML) takes data and searches for the patterns that underlie it. When it is finding something important to a particular issue, programmers will take their information and use it to evaluate different issues. What you need is the data that are reliable enough, and algorithms will distinguish valuable patterns. Data can come in the form of physical, satellite, visual, email, or unstructured data. ML is not a new field. We see that from the last 70 years, there are various computer scientists who have established many of their scientific and technical underpinnings. Nowadays, we talk about various innovations that can be paired with sense, for example, by collecting and manipulating images, sounds, and voice, computer vision, and audio processing, we can consciously interpret the world around them, for example, using facial recognition at kiosks on border control.

Comprehension is a concrete example of how AI can boost productivity: natural language processing engines and inference engines may require AI applications for the study and interpretation of knowledge gathered. Using this technology, search engine results are powered by the language-translation feature.

Countries are also substantially increasing the allocation of resources for Science, Technology, Engineering and Math (STEM) talent creation to create the future workforce for AI by investing in universities, mandating new courses (e.g., AI and law), and providing retraining schemes. For example, in the U.K., By 2025, over 1000 government-supported PhD researchers are scheduled to develop and set up a Turing fellowship to fund an initial cohort of AI fellows, while China launched a 5-year university program to train at least 500 teachers and 5000 students working on AI technologies. Structures of governance to allow all of the above mandates differ across countries. Many countries, such as the Ministry of AI (UAE) and the Office of AI and the AI Council (UK) have developed dedicated public offices, while China and Japan have allowed existing ministries to introduce AI in their sectoral areas.

Through commitments, such as rising R&D spending, setting up industrial and investment funds in AI start-ups, investing in networks and infrastructure, and AI-related public procurement, national governments have substantially increased public funding for AI. China, the USA, France, and Japan have devoted major public expenditure to the production and adoption of AI technology. These countries are also exploiting various public–private–academia combinations to grow and encourage AI. Developing technology parks and linking major companies with start-ups are some of the strategies (Discussion Paper National Artificial Intelligence Strategy 17). Some of the public-private collaboration approaches proposed by different national governments are the establishment of "national teams" with large private players to conduct fundamental and applied research. The architecture and applications of AI technology are rapidly evolving with significant implications for economies and societies.

A report by EY and NASCCOM found that by 2022, about 46% of the workforce will be engaged in completely new positions that do not exist today, or will be deployed in positions that have dramatically changed skill sets. It seems unlikely that any countries would be able to achieve and match the current momentum in the rapidly changing socioeconomic environment if they plan to wait a few years to develop an AI strategy and put in place the foundations for developing the AI ecosystem. The need of the hour is therefore to establish a policy structure to help set up a vibrant AI ecosystem in India.

We speculate in this chapter on how the AI can have an impact on the growth process. Our chief aim is to help shape the future research agenda. AI is not a revolutionary dream, but instead, one that is being incorporated with and implemented in several industries today. This covers areas, such as banking, national security, health care, criminal justice, infrastructure, and smart cities. There are several examples where AI is now affecting the environment and dramatically rising human capabilities. An AI program can or can act in the real world using technology, such as expert systems and inferencing engines. Examples of this in vehicles are autopilot features and aided-braking capability. The understanding and capability to learn and change by the passage of time underpins all three competencies. Many businesses already have a degree of AI, but the extent to which it is part of our everyday lives is likely to expand exponentially.

There are two main factors that make for AI growth:

1. Connection to infinite processing resources.
   In 2015, public cloud infrastructure was projected to cross almost US$ 70 bn worldwide. Data storage was common too.
2. Producing Big Data.
   The McKinsey Global Institute indicated that AI is contributing to the industrial revolution's transformation of society "ten times faster and at 300 times the size, or around 3000 times the impact." Computer capacities are expanding beyond routine work and some computer work is now being automated. Manufacturers are now starting to integrate ML into industrial machinery, gathering data to track and enhance the efficiency of manufacturing processes.

We need to think at the moment about what we want computers to do, how we want them to function, and how we are going to work with them.

8.3 AI STRATEGY FOR INDIA

According to studies, AI is expected to increase India's rate of innovation by 230% in just 2 years. The country has been focused on research in recent years, but it is still lagging behind world leaders.

A national AI plan, according to NITI Aayog's discussion paper on AI, "National Strategy for Artificial Intelligence," must be based on a framework that is tailored to the country's specific needs and goals. At the same time, it is capable of realizing India's full potential in terms of AI breakthroughs.

In May of this year, the NITI Aayog announced a proposal to establish an institutional framework in order to make AI a reality in India. It has issued a cabinet note to allocate Rs 7500 crores in funds, initially over 3 years, for the construction of a cloud computing platform called AIRAWAT and research institutes, as well as to establish a high-level taskforce to oversee the project.

In terms of economic impact, AI has the potential to fuel growth by automating multilayered physical tasks that demand adaptation and agility in a variety of industries. It also allows humans to focus on the roles in which they can provide the greatest value, enhancing human talents and increasing the capital efficiency.

8.4 SO, WHAT ABOUT THE FUTURE?

AI and robotics are anticipated to get integrated into many aspects of our daily life by 2025.

AI has been speculated to be invincible by 2040, with humans starting to lose both comprehension and power and human minds being increasingly taken out of decision-making processes.

This is extreme, I would say, as we cannot second guess where AI will lead, and since ML has evolved so much over the last decade, it is hard to determine what is going to be feasible in the future.

For instance, if machines, whose design has been taught by brain science, are able to think, predict, and react like the human brain, may the distinction between human and machine become blurred? It is important to note that there is a major difference between the ability to think (reason) and the ability to feel (sentience) when contemplating this issue, and although machines are becoming more intelligent, this does not mean that they do not have sentiments.

That said, even though they use sarcasm, idioms and emoticons, Rant and Rave (a consumer knowledge and interaction technology already in use) has a sentiment engine that understands what customers are thinking.

What is obvious is that there are going to be major technological changes quite quickly. In terms of human level intellect, however, it is unlikely that this will be accomplished any time soon, while surveys of leading AI researchers indicate a strong possibility of being accomplished in this century.

In the future, AI has the ability to result in "explosion" of intelligent robots that will work closely with humans in tandem with developments in robot hearing and vision.

International data have seen an annual growth rate of more than 50% compound annual growth rate (CAGR) since 2010 when more of the devices around us became related.

The computer science professor, told us: "Data is to AI what food is to humans."

And the rapid rise of data in an increasingly physical environment is continuously feeding AI upgrades.

With the introduction of technology in this digital era, we humans have pushed the boundaries of our thinking processes and are attempting to merge a natural brain with an artificial one. This on-going research spawned a brand new area called AI. It is the method by which a human can create a machine that is intelligent. AI is a subset of computer science that can recognize its environment and thrive in order to maximize its chances of success. AI should be able to do tasks based on previous experience. Deep learning, CNN, ANN, and ML are examples of fields that improve machine performance and aid in the development of more advanced technologies.

The term "Internet of Things" is defined as "thing to thing" communication. The system's three key goals are communication, automation, and cost savings. Dr. D.K. Sreekantha and Kavya A.M. discuss the use of IOT in agriculture and how it may benefit mankind.

8.5 THREE ROOMS OF ARTIFICIAL INTELLIGENCE-LED GROWTH

With AI as the latest development tool, it can boost growth in three essential ways or more. Firstly, it will build a modern automated workforce, what we call "intelligent automation." Secondly, AI will supplement and improve the current workplace and physical resource expertise and capabilities. Third, like other previous innovations, AI can accelerate economic growth. In time, this is a mechanism for large systemic change when markets that use AI are capable of coming out with creative activities and doing new things.

8.5.1 INTUITIVE MACHINING

Unlike traditional automation approaches, the latest AI-powered Smart Automation revolution now enables development across a variety of applications.

The first feature of this is the ability to automate dynamic, adaptable, and flexible physical world functions. This feature seeks to recover goods in a warehouse where businesses have focused on the ability of workers to navigate crowded conditions and avoid moving objects. Today, Fetch Robotics robots are using lasers and 3D depth cameras for safely handling and operating alongside warehouse staff. Within a modern facility, the machines can handle the vast majority of the products when used in combination with humans.

Although conventional automation technology is unique to the mission, the ability to solve challenges across sectors and work titles is the second defining characteristic of intelligent automation powered by AI. For example, Amelia-an IP soft AI program with the ability to interpret natural language has been supporting repair engineers in remote locations. Amelia will define a problem and suggest a response by reviewing all the books. The system has now studied the answers from mortgage lenders to the 120 most often asked questions, and has been used in a bank to manage these financial inquiries, usually a labor-intensive activity.

Works on the effect of AI, by Frey and Osborne (2013) and Author (2003), have historically been focused on the effects on employment, as certain jobs and tasks become automated and companies seek to create their own; business functions more efficiently. More recently, some authors focused on the benefits that could be derived from this.

The productivity gains that this automation brings, the future advantages and prospects of AI however go even more. The ability to capture, store, and interpret data on AI's size, speed, and the third and most powerful function of smart automation and self-learning is manner, which is permitted by scale repeatability. If Ms. X as a diligent employee acknowledges the issues in her expertise and takes steps to narrow them. When a question she can't respond is posed to Ms. X, she escalates it to an individual friend and then she watches how the person approaches the problem. The self-learning element of AI is a critical change. While conventional automation resources are degrading over time, smart automation assets are constantly improving.

8.5.2 MAJOR IMPACT ON LABOR AND CAPITAL

A significant portion of AI's economic development does not come from substituting current labor and resources, but from allowing them to be used even better. AI, for instance, will encourage humans to concentrate on aspects of their job that provides the most benefit. Workers at the hotel expend all of their time doing daily room supplies. More than 11,000 guest deliveries were made by the Relay fleet last year in the five major hotel chains where it is based. A Savioke CEO says: "Relay helps workers to focus their attention into can customer loyalty. AI also raises productivity by supplementing individual resources, giving workers new tools to improve their innate intelligence. Praedicat, for example, a firm that offers risk management tools to property and casualty insurers, is enhancing the risk-pricing capability of the underwriters. Learning by machine and to recognize significant potential threats, the AI website reads more than 22 million peer-reviewed scientific articles. As a consequence, not only do the underwriters take price risk but also create new insurance options, more precisely but still.[20]

AI can also boost resource efficiency, a vital element in industries in which it works. This represents an enormous sunk cost. For example, an industrial robotics firm Fanuc has partnered up with Cisco and other organizations in manufacturing to build a network to minimize plant downtime is estimated to cost US$ 20,000 a minute at one major car manufacturer. The Fanuc Intelligent Edge Link and Drive System (FIELD) is a framework for automation driven by advanced ML. It gathers and analyses data from the manufacturing cycle of different parts, intended to increase efficiency. In an 18-month "zero-shutdown" experiment, where it discovered significantly, FIELD was already introduced at one manufacturing industry. Likewise, we would expect more AI-intensive companies to (1) hire a greater proportion of (more highly paid) highly skilled workers, (2) outsource an increasing fraction of low-employment tasks, (3) give those low-employment workers a higher premium and keep within the company (unless we take the extreme view that the robots could perform all the functions to be performed by low-occupancy workers).[21]

8.5.3 SPREADING CREATIVITY

One of the AI's least-discussed benefits and the exponential data growth in an increasingly physical world continually feeds on AI upgrades is its ability to propel innovations as using a combination of lasers, global positioning systems, radar, cameras, computer vision, and ML algorithms, driverless vehicles can make it possible for a machine to feel and act accordingly in its environment. Not only are development giants joining the industry in Silicon Valley, but also mainstream firms do to build.

To continue to be relevant new partnerships, companies try to collaborate with them for future effect of driverless cars on economies and are likely to reach far beyond the automotive industry, as mobility engenders mobility. Mobile service providers will be expected to see a significantly higher demand from consumers as travelers, now able to enjoy leisure activities, such as driving, spend more time on the Internet, create new sales opportunities for service providers, and sell goods to their business partners. Product and service innovation could offer up to 7% or about $6 Billion potential GDP by 2030. AI can make a significant contribution by boosting innovation, which can then be used to improve current products and services, and new offers. The simulation suggests that innovation can contribute about 7%, leading to a possible $6 trillion increase in output by 2030, which is incremental to the output today. The first reason why these AI effects are significant is that companies can quickly improve their top lines to more efficiently enter underserved markets even with established goods and services.[22]

The value of the variable substitution gains depends on the efficiency gains that occur over time. The reason for the latter is that over the longer term, most technologies tend to encourage product innovation and services, boosting nontraditional industries and creating new markets altogether. Consider how the compact, high-pressure steam engine pushed outside the factory, leading to a rail and sea boom Voyage. In the early 1800s, the first steam-powered locomotive hits the rails in Britain and the first ship sailed until 1807 in the U.S. Consider how ICT was the basis for the internet economy, now reshaping the retail, transport, and media sectors.

8.6 CLEARING THE PATH: FOR THE FUTURE OF AI

Various entrepreneurs have cautioned that AI might become "the greatest existential threat" for mankind. Futurist Ray Kurzweil's more positive opinion is that AI will help us make "huge moves to tackle the [world's] great challenges." The truth is, all of these depend on how we manage the transition to an AI era.

8.6.1 PREPARING THE NEXT GENERATION FOR THE AI FUTURE

More important would be the effective convergence of human intelligence with AI, so that they create a learning which has an understanding of both the areas effectively as never before. For the separation of duties between connections with machine and man, policymakers need to re-evaluate the kind of information and skills that should be imparted to future generations. Technical education is currently moving in one direction. People are learning how to use computers. This will bring a major change, as machines learn from humans, and people benefit from computers to meet AI's pledge as a new development force that can rekindle the economic growth, and related stakeholders need to be fully prepared scientifically, technically, economically, ethical and social connections in order to tackle the challenges which AI becomes the most important aspect of our lives, it emerges. The area of departure understands complexities.[23]

8.6.2 FOSTER AI-POWERED CONTROL

When autonomous computers perform roles that have been performed solely by humans, existing rules, it needs revisiting. New York Law of 1967 requiring drivers to keep one hand on the steering wheel was meant not only to improve safety but also to obstruct semiautonomous safety innovations like automatic lane centralization.

In other cases, it calls for new regulation. For starters, although AI may be tremendously helpful in making medical diagnoses, doctors resist using such innovations, believing that they would be exposed to allegations of malpractice. This confusion could discourage and even delay the take-up of AI.

8.6.3 DEFEND AN AI ETHICAL CODE

Intelligent species are emerging quickly into social environments that were once only populated by humans. This brings up legal and social issues that can delay AI's growth. This varies from how to respond to racially discriminatory algorithms to how autonomous vehicles in the case of an accident will offer their driver's life priority over others. Seeing how pervasive smart devices can be in the future, policymakers need to maintain an AI code of ethics. More concrete standards and best practices in the production of smart devices will accompany ethical debates.

8.7 COUNTER THE CONSEQUENCES OF REDISTRIBUTION

Many commentators worry that AI can cut down jobs, exacerbate inequality, and erode incomes. This describes the increase of global marches and debates of places like Switzerland on the adoption of simple guaranteed pay. Policymakers must understand the importance of such apprehensions. Their answer is expected to be equal. First of all, policy-makers will emphasize how AI will offer real benefits. For example, AI can increase employee satisfaction. A survey tells us that almost 84% of managers expect that robots can make them more productive and more important for their jobs.

Outside the office, AI promises to solve some of the toughest challenges in the world, such as changes in climate and inadequate health care coverage (by reducing the pressure on overwhelmed systems). These benefits should be explicitly expressed to promote a more optimistic view of potential AI. Second, lawmakers must consciously counter and avoid the downsides by AI. Such reforms would negatively impact other classes. To avoid a backlash, policymakers should find out new groups at a very high risk of change and should create various policies which can be added into areas of economics.[24]

8.8 AI MAY HELP PROGRESS TOWARDS SDGs

In 2015, the 17 Sustainable Development Goals (SDGs) were adopted by the 193 UN Member States, It contains 169 concrete goals and tackles the most urgent needs of the world, with the goal to be completed by 2030. The UN defines "sustainable development" as financial prosperity, social growth, and environmental conservation. AI has great potential to promote sustainable development in each of these countries' expertise in the economy. Adding AI to hunger eradication efforts could help to increase production and profitability in fields, such as forestry, milk, and other logistics. AI researchers and universities integrate AI, robots, and sensor technology to enhance plant breeding and farm management. This project focuses on crop varieties—such as sorghum, a heat-tolerated grain that is grown in Africa, in developed countries where it is the most desired economic expertise. Adding AI to an effort to eradicate hunger could help. AI can also be used to build more reliable measures where deprivation is spread and how it spreads quickly and helps to decide where money could be distributed. For example, a US research team overlays high-resolution night images, daytime images, and the use of ML methods for calculating the energy income and asset resources of African nations. Using AI in transportation is becoming commonplace, especially for ensuring safety. Smart sensor technology is expected to have AI to help improve current track signaling. It is an AI-controlled pedestrian crossing that uses numerous cameras and neural networks to track cyclists, cars, and other moving objects, interprets their movement pattern and provides simple LED warning signals to motorists in real-time.[25]

8.9 CORPORATE FUNDING

AI will allow all students to access new opportunities and quality education no matter what circumstances the student has inherited. AI majorly uses AI-enabled technologies in the area of education, including learning-specific technologies evaluations or specific improvements to the tasks. AI takes advantage of high-resolution satellite imaging and deep learning technologies in an endeavor to map at all the colleges and schools everywhere in the world. AI is also supporting to enhance and increase the change toward equality and toward gender through the creation of goals and effective ways of identifying and addressing gender bias, gender discrimination, and violence. Textio's statistical database, for example, cites data from more than 45 million workers posts and recruiting reports to gage career posts gender tone. The index attracts attention to innocuous words like "brave," "working" and "strengthening." Proven to build more male workers even as wearable fitness tracking can enhance healthier practices, the AI-based financial planner will track users' financial vitalities to boost their financial status. For example, research has demonstrated how an AI-powered consultancy functions can give customized advice to individuals. Knowing from historical facts, the advisor can forecast individual's account balances for a future date, classify recurring accounts, budget costs, can calculate unnecessary major expenditures and analyzes the budget, analyzes the spending nature of the customer group by category. Changes are minor, but important, will add up and empower the politically vulnerable AI research has undergone a tradition of positive waves accompanied by skepticism and frustration. The Inertia Periods is also called "AI Winters." Any previous breakthrough has only partly lived up to the hype that it developed and none failed to kick-start the mainstream technology. Protect the climate. There are ways to boost handling efficiently. Globally, work into autonomous vehicles has spurred development, especially in computer AI fields and robotics and vision. Due to the extremely high market potential, most of the large companies in the last 2 years have invested in AI especially in the autonomous vehicle industry, since it is commonly considered as the first technical use of AI and will be implemented on a wide scale. Besides, AI algorithms are based on Indian roads due to congestion and turbulent situations in Indian traffic. Driving data can be highly reliable. According to the Stanford AI Index, error rates in object recognition have fallen from 28.5% to 2.5% since 2010[26] for natural capital and responsibility for dangerous practices by the use of AI. For example, Pluto is an application of AI analytics that uses advanced water treatment sensors saving time, energy, and water. The framework uses a combination of supervised and uncontrolled learning to analyze data on results from the structured and unstructured reams. AI research has undergone a tradition of positive waves accompanied by skepticism and frustration—Inertia Periods, called "AI Winters." Any previous breakthrough has only partly lived up to the hype that it developed and none failed to kick-start the mainstream technology.

8.10 RAPID ACCEPTANCE OF AI

This year, as signs emerge that the great disruption of AI employment would be a false one.

People are more likely to consider AI in the workplace and more readily in culture. We might hear less about robots taking our jobs and more about robots creating new jobs. Our jobs are made easier by robots. In turn, that can lead to a faster adoption of AI as expected by some organizations.

8.11 ORGANIZATIONAL RETOOLING WILL COMMENCE

It is going to be a lengthy process, but some organizations that are forward thinking are trying to reduce the silos that separate data into cartels which are already breaking down the staff into isolated units. Some are also going to begin on the huge up skilling that AI and other emerging technologies need. It will teach a new mindset that emphasizes collaboration with co-workers and with AI (Figs. 8.1 and 8.2).

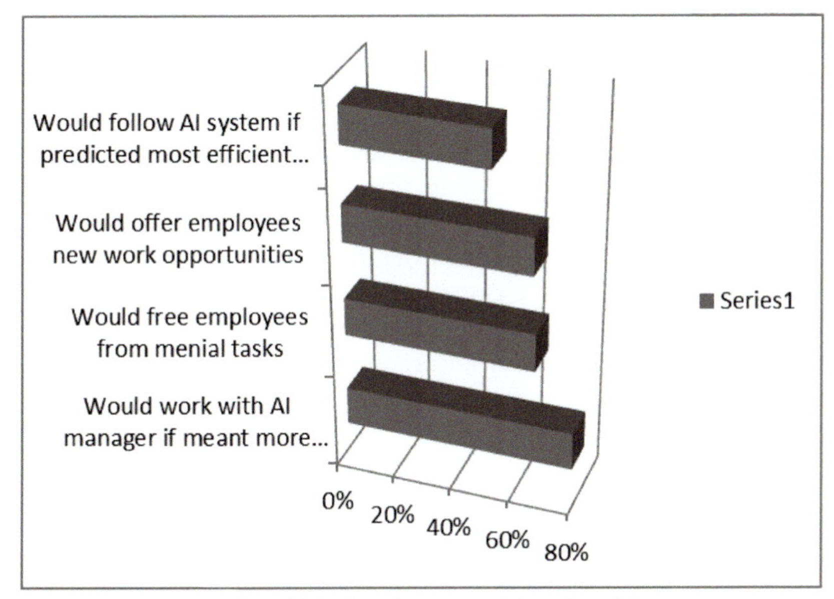

FIGURE 8.1 How workers think about human machine AI models.

In spite of such benefits, there are certain reasons which are holding the impact of AI in enterprises.

They are as follows:

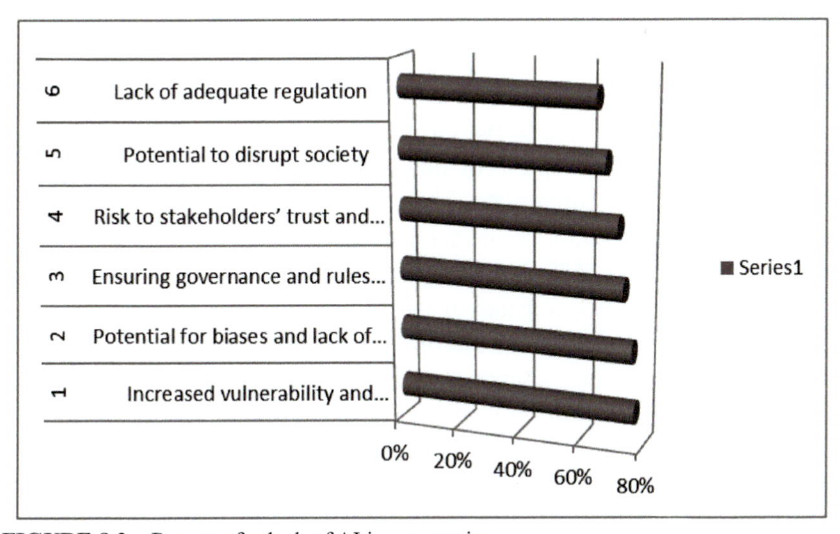

FIGURE 8.2 Reasons for lack of AI in companies.

8.12 INDIAN AI PLAN

According to research, within only 2 years, AI is expected to increase the rate of creativity within India by around 230%. In recent years, the nation has concentrated on the study, but is still far from keeping pace with the world leaders. According to NITI Aayog's discussion paper on AI's "National Strategy for Artificial Intelligence," a regional AI plan needs to be based on a framework that is customized to the unique needs and aspirations of the country and at the same time being able to recognize India's maximum potential to harness developments in different fields.[27]Universally, corporate leaders and investors believe that AI and ML can change their companies through cost savings, risk control, streamlining processes, stimulating growth, and driving creativity. AI has tremendous potential to fuel sales and income growth. Marketing, customer care, and distribution were listed as the top three roles where according to a Forbes study of 1093 executives, AI would achieve its full potential.[28]

Research and Markets.com's offering has been added to the "AI Global Market Report 2020–2030: COVID-19 Development and Transition" report. This report offers the essential details that strategists, marketers, and senior management need to analyze the global demand for AI.

The focus of this report is on the demand for AI that is experiencing strong growth. The report offers a roadmap to the demand for AI that will influence and transform our lives over the next 10 years and beyond, including the reaction of the markets to the threat of the global pandemic at a CAGR of 43.39%, the global AI market is projected to grow from $28.42 billion in 2019 to $40.74 billion in 2020.

The rise is mainly due to the global health emergency of COVID-19, which has led to a new wave of innovative technologies emerging as a potential solution to contain the epidemic, including disruptive AI technology (e.g., smart machines and robots). The business is then projected to rebound and hit a 34.86% CAGR of $99.94 billion in 2023.

The demand for AI consists of the selling of applications for AI and associated services. Computer intelligence is also classified as AI. AI is a large branch of computer science associated with the development of smart machines capable of performing tasks requiring human intelligence.[29]

In 2019, the largest area in the AI industry was North America. In the forecast period, Asia-Pacific is expected to be the fastest growing region.

Apple purchased xnor.ai, a Seattle-based company, for $200 million in January 2020. In providing AI capabilities, the acquisition is expected to reinforce Apple. Xnor.ai is listed in Forbes AI 50 2019, the most promising AI company in America.

By offering hardware, software services, the AI market covered in this report is segmented. It is also divided into ML technology, natural language processing, context-aware computing, computer vision, others (image processing, speech recognition) and health care, automotive, agriculture, retail, marketing, telecommunications, defence, aerospace, media and entertainment, and others by end-user industries.

The small number of AI experts in the AI industry is a major challenge. The organization needs AI experts, professionals, and researchers to create AI software. There are 300,000 AI researchers and practitioners worldwide, according to the study report in 2017 by Tencent, a Chinese tech giant, but there is demand for about millions of positions in the industry . The growth of the AI market may be affected by the limited number of AI experts.[30]

Many businesses are introducing Advanced ML. Implementing conventional ML models for real-world business problems is very difficult. Automated ML, which helps non-ML practitioners to use ML algorithms without becoming ML wizards, is the better option. For example, tools such as "Google cloud AutoML" are used with minimum ML expertise to train custom-made and high-quality ML models.

The growing investment in AI technologies has led to the growth of the market for AI. In order to improve the efficacy of AI software, many businesses are investing in AI technology or AU start-ups, as AI helps them to make more informed decisions and produce better results. Microsoft, for example, has invested about $1 billion in a San Francisco business called Open AI. A collaboration was formed between the two companies to develop AI supercomputing technologies on Microsoft's Azure cloud. AI is beneficial across various industries. AI will, for example, contribute to breakthroughs in the health care sector in R&D and diagnostics. It will help to make better predictions in banking, manufacturing, and improve productivity and production controls. Therefore, because of its advantages in every field, the increase in investment in AI technologies is driving the growth of the AI industry.

One of the tendencies driven by the epidemic was the usage of automation. AI is not only being adopted by corporations. Economies are also focusing on improving their AI skills as a means of boosting growth. Developed economies, such as the United States, China, and the European

Union are already competing. India is now likely to join them. In recent years, the Indian government established an AI Task Force and directed the NITI Aayog to develop a national policy on AI with the goal of leveraging AI for inclusive growth. In India, however, AI adoption is still in its infancy. Responsible AI for Social Empowerment (RAISE) 2020, India's inaugural AI summit, has turned the possibility of AI adoption into a near-term reality.

India appears to be on the verge of the fourth industrial revolution, thanks to the rapid adoption of AI-based technologies. Low-cost labor may lose its competitive edge in the near future as countries begin to reap the benefits of AI in the form of higher productivity and cost advantages, making it more profitable than labor. As a result, it would be prudent for India to strengthen its AI skills, lest the global digital divide grow any further and leave us behind.

8.13 CONCLUSION

Established countries tend to find other ways to increase their productivity, including best practices and business restructuring. Therefore, they could have less reason to press for AI (which may give them a comparatively lower economic gain than developed economies do in either case). Many developing countries are possible exceptions to this law. China, for example, has a national strategy in place and is investing heavily to become a world leader in the supply chain for AI. Where this technology could create new jobs in the future is difficult to determine, yet it is easier to see which tasks AI might take from humans. Any routine, repeatable task is likely to be automated. Since decades, this change toward automation has occurred, but what is new now is that it affects far more industries. It is possible that by inventing completely different forms of jobs and taking advantage of our specific human capacities, we can respond to technical changes. Since many countries face a lot of internal opposition today, there is a problem for all of them to follow policies that are viewed as welcoming not just to their national's audience but still their mates from around the world. Consequently, there is a lot of rivalry for the same constrained space. Although there is a major problem of employment, health metrics, infrastructure, corruption, governance, etc. are problems between India and other countries to find and get sufficient financial and technical assistance from these nations.

Given the challenges, India can cooperate with some of the countries picked. This could draw the right set of attention for improving country's quality education and health. Given that the country is yet to bring broad infrastructure of highways, railways, housing, urban development, etc., India could draw the technical and financial support of the developed countries. India could improve its exports and achieve faster growth by better cooperation and agreements with advanced countries and by curbing its own protectionist tendencies.

The planet, through AI and data analytics, is on the verge of revolutionizing several industries. There are now major implementations that have altered decision-making, business models, risk reduction, and system efficiency in banking, national security, health care, criminal justice, transport, and smart cities. Such technologies produce tremendous economic and social benefits.

Yet, there are significant consequences for society as a whole for the way in which AI systems unfold. It matters how policy problems are handled, ethical disputes are reconciled, legal realities are resolved, and how much accountability is needed in AI and data analytical solutions. The way in which decisions are taken and the way in which they are implemented into organizational routines are influenced by human choices about software creation.

It is important to better understand just how these procedures are carried out, because they will soon have a huge effect on the general public and for the near future. In human affairs, AI could well be a revolution and become the single most influential human invention.

KEYWORDS

- **supply chain**
- **artificial intelligence**
- **growth**
- **business**
- **technology**
- **gross domestic product**

REFERENCES

1. Goel, R. Moving from Cash to Cashless Economy: A Study of Consumer Perception towards Digital Transactions. *Int. J. Recent Technol. Eng.* May **2019,** *8* (1), 1220–1226.
2. Gupta, A. An Analysis of Financial Fraud through PNB Bank Scam and Its Technical Implications. In *2020 International Conference on Computation, Automation and Knowledge Management*; IEEE: Noida, 2020.
3. Gupta, A. Application of Artificial Intelligence for Sustaining Green Human Resource Management. In *2018 International Conference on Automation and Computational Engineering*; IEEE: Noida, 2018.
4. Puri, N. Mobile Banking a Myth or Misconception. In *2020 10th International Conference on Cloud Computing, Data Science & Engineering*; IEEE, 2020.
5. Rani, S. Consumer Adoption of Smart Biometric Lock Among SAARC Nations. *Int. J. Eng. Adv. Technol.* **2019,** *8* (4C), 138–144.
6. Sahai, S. Impact of Digital Commerce on Fashion Industry to Gain Customer Loyalty. *Int. J. Eng. Adv. Technol.* **2019,** *8* (5), 730–740.
7. Sahai, S. Impact of Digitization on Impulse Buying—What Makes the Customer Bite the Bait. *Int. J. Innov. Technol. Explor. Eng.* May **2019,** *8* (7), 2948–2952.
8. Singhal, A. A Study on Transformation in Technological Based Biometrics Attendance System: Human Resource Management Practice. In *2018 8th International Conference on Cloud Computing, Data Science & Engineering*; IEEE: Noida, 2018.
9. Tiwari, P. A Study of Consumer Adoption of Digital Wallet Special Reference to NCR. In *2019 9th International Conference on Cloud Computing, Data Science & Engineering*; IEEE: Noida, 2019.
10. Tiwari, P. Application of Artificial Intelligence on Behavioral Finance. In *Recent Advances in Intelligent Information Systems and Applied Mathematics*; Castillo,O. J., Ed.;. Springer, 2020.
11. Tiwari, P. Linkage between Supply Chain Operation Reference model and Learning Style Diversity—An Empirical Study in Indian Logistics Industry. *Int. J. Supply Chain Manage.* **2019,** 1025–1032.
12. Tiwari, P. The Study to Analyze the Impact of Green Supply Chain Management in India. *Int. J. Supply Chain Manage.* **2019,** *8* (3), 1033–1038.
13. Tomar, K. S. A Study of Business Performance Management in Special Reference to Automobile Industry. In *Data Management, Analytics and Innovation*; Sharma, N. C., Ed.; Springer, 2019.
14. Venaik, A. Information Security Parameters Used By Aadhar, Uidai And It's Impact. *Int. J. Sci. Technol. Res* **2019,** *8* (10), 1150–1154.
15. Vinaik, A. The Study of Interest of Consumers In Mobile Food Ordering Apps. *Int. J. Recent Technol. Eng.* **2019,** *8* (1), 3424–3429.
16. https://niti.gov.in/writereaddata/files/document_publication/NationalStrategy-for-AI-Discussion-Paper.pdf
17. https://www.globenewswire.com/news-release/2020/06/04/2043624/0/en/Global-Artificial-Intelligence-Market-Report-2020-to-2030-COVID-19-Growth-and-Change.html

18. https://www.brookings.edu/research/how-artificial-intelligence-is-transforming-the-world/
19. https://www.shponline.co.uk/leadership-and-innovation/artificial-intelligence-current-trends- and-future-developments/
20. Srivastav, S.; Garg, V.; Gupta, A. 'Bridging The Perceived Gap between Industry and Academia'. *Int. J. Supply Chain Oper. Resil.* **2020,** *4* (2), 202–216.
21. Sawhney, S.; Kacker, K.; Jain, S.; Singh, S. N.; Garg, R. "Real-Time Smart Attendance System Using Face Recognition Techniques". In *2019 9th International Conference on Cloud Computing, Data Science & Engineering (Confluence)*; Noida, India, 2019; pp 522–525. doi: 10.1109/CONFLUENCE.2019.8776934
22. Hardaha, P. N.; Singh, S. "Data Mashup for Improving the Performance of Global Network". In *2019 9th International Conference on Cloud Computing, Data Science & Engineering (Confluence)*; Noida, India, 2019; pp 197–202. doi: 10.1109/CONFLUENCE.2019.8776906
23. Kumar, A.; Kumar, P. S.; Agarwal, R. "A Face Recognition Method in the IoT for Security Appliances in Smart Homes, Offices and Cities". In *2019 3rd International Conference on Computing Methodologies and Communication (ICCMC)*; Erode, India, 2019; pp 964–968. doi: 10.1109/ICCMC.2019.8819790
24. Agarwal, R.; Sharma, H.; Sharma, N. "Exponential Scale-Factor Based Differential Evolution Algorithm". In *2017 International Conference on Computer, Communications and Electronics (Comptelix)*; Jaipur, India, 2017; pp 354–359. doi: 10.1109/COMPTELIX.2017.8003993
25. Tyagi, L.; Singhal, A. "Neuro-Fuzzy Approach to Explosion Consequence Analysis". In *2020 10th International Conference on Cloud Computing, Data Science & Engineering (Confluence)*; Noida, India, 2020; pp 315–319. doi: 10.1109/Confluence47617.2020.9058024
26. Gupta, S.; Seth, S.; Dhawan, A.; Singhal, A. "Location Based Camera Disable System on Android Platform". In *2018 8th International Conference on Cloud Computing, Data Science & Engineering (Confluence)*; Noida, India, 2018; pp 14–15. doi: 10.1109/CONFLUENCE.2018.8442511
27. Sharma, P.; Zhang, J.; Ni, K.; Datta, S. "Time-Resolved Measurement of Negative Capacitance". *IEEE Electr. Device Lett. Feb 2018*, *39* (2), 272–275. doi: 10.1109/LED.2017.2782261
28. https://www.analyticsinsight.net/artificial-intelligence-india-comprehensive-overview/
29. https://www.sciencedirect.com/science/article/pii/S2589721719300182
30. https://economictimes.indiatimes.com/news/economy/policy/view-artificial-intelligence-for-inclusive-growth/articleshow/78566916.cms?from=mdr
31. https://www.pwc.com/us/en/advisory-services/assets/ai-predictions-2018-report.pdf

CHAPTER 9

Monitoring System for Greenhouse Using a Deep Learning Technique

ADITYA KAKDE[1], NITIN ARORA[2*], DURGANSH SHARMA[3], MAMTA MARTOLIA[4], and ALANKNANDA ASHOK[5]

[1]*Department of Systemics, School of Computer Science, University of Petroleum and Energy Studies, Dehradun, Uttarakhand, India*

[2]*Department of Informatics, School of Computer Science, University of Petroleum and Energy Studies, Dehradun, Uttarakhand, India*

[3]*Department of Cybernetics, School of Computer Science, University of Petroleum and Energy Studies, Dehradun, Uttarakhand, India*

[4]*Department of Computer Science and Engineering, Uttarakhand Technical University, Dehradun, Uttarakhand, India*

[5]*Department of Electrical Engineering, Women Institute of Technology, Dehradun, Uttarakhand, India*

[*]*Corresponding author. E-mail: narora@ddn.upes.ac.in*

ABSTRACT

Monitoring of plants plays a vital role in greenhouse as with the growth, it is also important to check if the plants are not affected by the diseases as fresh fruits and vegetables can only grow in a healthy region. Monitoring each and every plant is very time consuming and hard working. There are also chances that some plants can slip from under the eye of monitoring and can destroy the plantation especially in case of plants such as tomato. Thus, an automated monitoring process is needed to monitor the health and type of plants. This chapter proposes an architecture which will help to monitor the plants by taking the images in real time and then classify it. The model implemented for this classification is based on convolutional neural network, and a novel approach has been used to get better performance for classification task. The check of performance has been done on the basis of four parameters that are train loss, train accuracy, test loss, and test accuracy.

9.1 INTRODUCTION

Greenhouse is a structure that helps us to grow any type of crops without keeping in mind the changes of seasons. Varying in size, it can be from prototype to industrial-sized buildings. The miniature one is known as a cold frame. The structure is mainly covered with transparent material like glass so that no insects and pests can enter inside.

The idea basically came from the time of the Roman Empire. A Roman emperor named Tiberius liked to eat cucumber-like vegetable daily. Thus, the gardeners used to grow it using an artificial method in which they planted that crop in wheel carts that were used to be put in the sun daily and then taken inside at night to keep them warm. After that, the crops were stored in the frames that were used to be covered with oiled cloth also known as specularia.

FIGURE 9.1 Left figure shows the complete automation of greenhouse, and right figure shows a diseased leaf.[15,16]

Many different methods for automation were proposed like glowing LEDs inside the greenhouse to provide photosynthesis whenever required, maintaining the moisture and humidity of air by using fogger, estimation of volume of the storage containers (vegetables and fruits) using different types of sensors, giving an alert about the empty water tank, and so on. These automation processes are done so that we can get vegetables and fruits in unseasonable time. But these all processes help only in the growth of plants. During the period of evolution from infant to adult, there are chances that they can also be affected with some diseases. Thus, monitoring to stop the growth of disease is a very crucial task. Neglecting it can surely affect the growth of plants and overall decrease in the quality of vegetables and fruits. This not only affects plants but can also prove to be a complete waste of the above processes because if there are no healthy plants, there is no meaning of automation.

For the proposed automated monitoring part, a famous deep learning network is used which is named as convolutional neural network (CNN). It was proposed by LeCun and Bengio,[1] inspired by the mammal's visual cortex.

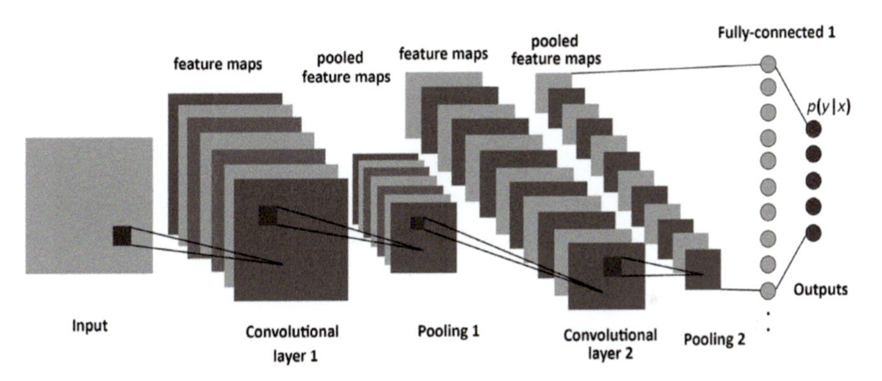

FIGURE 9.2 Stacking of convolutional and pooling layer followed by fully connected layer.

Source: Reprinted from Ref. [17].

It has the ability to capture the most important features from the images. First, it uses the concept of receptive field due to which it is able to scan all the images and then extract those features. Then, the next process is governed by the convolutional process which means the method of dot product between the values of filters and the input image. It also consists of some other parameters that are number of filters, strides, and padding. Strides specify the distance between the convolutional layer and the input image or can be said as the length of the receptive field. Padding comprises two categories that are valid and same. Valid means not be kept the same dimension of the input and output image. Same means to keep the same dimension of input and output image. Before passing through convolutional layer, it has to pass through activation function so that only the sensible information can be forwarded. After this, it is passed to pooling layer for down-sampling and then to fully connected layer for the classification task.

9.2 RELATED WORK

9.2.1 MULTILAYER PERCEPTRON AND CONVOLUTIONAL NEURAL NETWORK

LeCun and Bengio[1] presented a paper, *Convolutional Networks for Images, Speech and Time-Series*, where they proposed that there is no need of hand-crafted feature extraction method if CNN is used for image recognition. But there is a need to normalize the image for size and orientation. It is also specified that fully invariant recognition is still beyond reach. Gu et al.[2] presented a paper, *Recent Advances in Convolutional Neural Networks*, in which they have provided a broad survey on improvements in CNN in the past few years which describes pooling layer, loss, weight initializations, regularizations, optimizations, and gradient descent. Further, stress was given on describing fast processing of CNNs and application of CNNs.

Ramchoun et al.[3] presented a paper, *Multilayer Perceptron: Architecture Optimization and Training*, where they applied genetic algorithm on multilayer perceptron to optimize the training. The multilayer perceptron consists of two hidden layers with sigmoid activation function, and for the training purpose back-propagation algorithm is used. The proposed network is tested on Iris dataset where it has given the highest training accuracy of 98.7%, when compared with other machine learning models like SVM, RBF, EBP, and MLP. Kakde et al.[4] presented a paper, *Novel Approach towards Optimal Classification using Multilayer Perceptron*, where it was proposed of using a particular number of hidden layers with best possible activation function. Thus, multilayer perceptron having three hidden layers with exponential linear unit (ELU) activation function has given the best average results that are 0.014652 and 0.990671 when compared to 0 and 1. But when the alternate values are found smaller and greater, further investigation was done where it was found that the loss decreased much faster for MLP with three hidden layers with ELU activation function that is 0.00429 at 600 epochs. In addition, 0.5 learning rate is mentioned which tends to be best suited for the classification task on XOR operation.

9.2.2 ACTIVATION FUNCTIONS

Clevert et al.[5] presented a paper, *Fast and Accurate Deep Network Learning by Exponential Linear Unit (ELUs)*, where they proposed a new activation function called as ELU which is capable to take negative inputs. When tested on MNIST dataset, it yields the test error of ±0.24%. In addition, when compared with other networks such as Alexnet, DSN, NiN, Maxout, All-CNN, Highway Network, and Fract. Max-Pooling on CIFAR-10 and CIFAR-100 dataset, the proposed network gave the least test error of 27.62% on CIFAR-100. For CIFAR-10, Fract. Max-Pooling achieved the lower error rate of 4.50%. Dabal Pedamonti[6] presented a paper, *Comparison of Non-linear Activation Functions for Deep Neural Networks on MNIST Classification task*, where four different activation function which are ReLU, LReLU, ELU, and SELU were compared and were tested on the basis of loss and accuracy when used MNIST dataset. During this test, ELU outperformed all other activation function $2.05e^4$ training loss at 0.1 learning rate, $1.15e^1$ testing loss at 0.1 learning rate, 1.000 training accuracy at 0.05, 0.1, and 0.2 learning rate, and 0.981 testing accuracy at 0.2 learning rate.

Shang et al.[7] presented a paper, *Understanding and Improving Convolutional Neural Networks via Concatenated Rectified Linear Units* in which a new activation function was proposed and was implemented on different neural network models based on convolution. These networks were then tested on CIFAR-10/100 and ImageNet dataset. When tested on CIFAR-10, CReLU, and CReLU, half of the single test gave 8.43 and 8.37 error rate, on average gave 9.39 ± 0.11 and 9.44 ± 0.09 error rate, and on vote gave 7.09 error rate. When tested on CIFAR-100 on single test gave 31.48 and 33.68 error rate, on average gave 33.76 ± 0.12 and 36.20 ± 0.18 error rate, and on vote, gave 27.60 and 29.93 error rate. On VGG network, when CReLU applied on conv 1,3 and tested on CIFAR-10, gave the least error rate of 5.94 in single, 6.45 ± 0.02 in average, and 5.09 in vote. When applied on conv 1,3,5, gave the least error of 26.16 in single, 27.67 ± 0.07 in average, and 23.66 in vote. Further, VGG + CReLU were compared with other neural network models in which it has achieved the least error rate of 5.09 in CIFAR-10 and 23.66 in CIFAR-100 datasets.

When tested on ImageNet dataset, conv 1–4 achieved the least top 1 and top 5 error rate of 39.82 and 18.28 and conv 1,4,7 achieved least top 1 and top 5 error rate of 35.70 and 15.32. Further comparison was done with other neural network models where conv 1,4 achieved the least top 1 and top 5 error rate of 39.82 and 18.28 whereas conv 1,4,7 achieved least top 1 and top 5 error rate of 35.70 and 15.32.

9.2.3 OPTIMIZATION

Ioffe and Szegedy[8] presented a paper, *Batch Normalization: Accelerating Deep Network Training by Reducing Internal Covariate Shift* in which a concept of batch normalization when combined with Google Inception Model outperformed existing models that are GoogleNet Ensemble, DeepImage Low-res, DeepImage high-res, DeepImage Ensemble, BN-Inception Single Crop, and BN-Inception Multicrop. BN-Inception ensemble achieved the lowest top 5 error of 4.9% and BN_X30 also called batch normalization with inception at 0.045 learning rate achieved the maximum accuracy of 74.8% on ImageNet Dataset.

9.2.4 GREENHOUSE MONITORING AND AUTOMATION

Kamilaries et al.[9] presented a paper, *Deep Learning in Agriculture: A Survey*, where a survey on 40 research papers was presented in which latest techniques of CNN are discussed. Further application of deep learning in agriculture was discussed which consists of areas of use, data sources, data variations, data preprocessing, data-augmentation, technical details, outputs, performance metrics, overall performance, generalizations on different datasets, and performance comparison with other approaches. In addition, advantages and disadvantages of deep learning and the future of deep learning in agriculture were discussed.

Rodriguez et al.[10] presented a paper, *A System for the Monitoring and Predicting of Data in Precision Agriculture in a Rose Greenhouse Based on Wireless Sensor Network*, in which a wireless sensor has been integrated to capture the data such as temperature, humidity, and light and then data mining techniques sugt ch as linear regression, neural network, and SVM were implemented on it. A relative absolute error of 13.18% was obtained when used linear regression. After that, when used neural network gave a high mean absolute error of 0.998°C. Further, SVM was implemented which gave a relative absolute error of 11.33% and was considered best among other two.

Gang et al.[11] presented a paper, *An Agricultural Monitoring System Based on Wireless Sensor and Depth Learning Algorithm*, in which a novel method called ZigBee was implemented, a wireless sensor network technology which is used to measure soil moisture and humidity with low-power consumption and cheap cost. After that, deep learning model called as automatic encoder was used to accurately estimate moisture content in soil and auto-control the operation of irrigation system.

9.3 METHODOLOGY

9.3.1 ACTIVATION FUNCTIONS

Activation function can also be called transfer functions. It range lies between 0 and 1, −1 and 1 or depending upon the property of that particular activation function. They are normally used to produce a sensible output. They introduce nonlinear functional mapping to the network. Inner activation function and outer activation function are the two categories which come under this domain. The inner activation function means those which can be used with hidden layers like linear, tanh, sigmoid, ReLU, ELU, and so on, whereas outer activation function means those which can be used with the output layer like linear, sigmoid, and softmax.

9.3.1.1 EXPONENTIAL LINEAR UNIT

ELU() activation function can also be called as ELU which is described as

$$\text{elu}(x) = \begin{cases} \alpha\left(e^x - 1\right), & x < 0 \\ x, & x \geq 0 \end{cases} \tag{9.1}$$

and, its gradient is described as

$$\frac{\mathrm{d}\,elu(x)}{\mathrm{d}x} = \alpha\left(e^x - 1\right) + \alpha \tag{9.2}$$

where α is a constant whose value is denoted as 1. For the positive value of x, it acts as an ReLU but unlike ReLU, it also accepts negative inputs. As in ReLU, if we input any negative value, it is then automatically converted to 0, and because of this, ReLU results in dead neuron. When talking about ELU, the values are saturated in the negative part of the domain whose range lies in between −1 and 0. The mean activation comes closer to zero which increases the learning of the network. When talking about positive inputs, the mean activation in ELU is closer to 0 than in ReLU. As ReLU have higher value of mean activation then ELU, it introduces a bias in the next layer which slows down the learning. Every mean activation introduces a bias but as mean activation is directly proportional to the bias, thus ELU beats ReLU in positive inputs too in terms of learning. Thus, this chapter also used ELU activation function on first two fully connected layers and in highway layer.

9.3.1.2 CONCATENATED RECTIFIED LINEAR UNIT

Concatenated rectified linear unit can also be termed as CReLU like ReLU, it takes both negative as well as positive inputs. In ReLU, when a positive input is entered, the gradient becomes 1 and if negative input is entered, the gradient becomes 0, thus resulting in dead neurons. As CReLU take both type of inputs thus consists one positive ReLU and one negative ReLU. These two then concatenates to form a concatenated ReLU. It means for positive x, CReLU outputs $[x, 0]$ and for negative x, it outputs $[0, x]$, thus increasing the output dimension. Due to this regularization, invariance and reconstruction properties are also improved. Mathematically, it can be given as

$$\text{CReLU}(x) = \left[\max(0, x); \quad \max(0, -x)\right] \tag{9.3}$$

CReLU activation function is used in the last hidden layer of the network as it increases the output dimension, but it was not used in the starting layers as it slows down the training of the network.

9.3.2 BATCH NORMALIZATION

Batch normalization is the normalization of output of the hidden layer before passing from any activation function of a particular hidden layer. Batch normalization reduces the problem of vanishing gradient by giving freedom to the hidden layers to learn some of the features of the input by itself. It consists of two parameters that are α and β. These two parameters are learned by the network. Because of these two parameters, mean and variance of the network are fixed. Due to this, the hidden layer of the network does not have to be much dependent on the previous layers. In addition, the oscillation which occurs during the gradient descent process is reduced while approaching toward global minimum. The formula can be given as

$$\mu_\beta \leftarrow \frac{1}{m}\sum_{i=1}^{m} x_i \quad \text{Mini batch mean} \tag{9.4}$$

$$\sigma_\beta^2 \leftarrow \frac{1}{m}\sum_{i=1}^{m} \left(x_i - \mu_\beta\right)^2 \quad \text{Mini batch variance} \tag{9.5}$$

$$x_i \leftarrow \frac{x_i - \mu_\beta}{\sqrt{\sigma_\beta^2 + \epsilon}} \quad \text{Normalize} \tag{9.6}$$

$$y_i \leftarrow \gamma x_i + \beta \equiv \mathrm{BN}_{\gamma,\beta}\left(x_i\right) \quad \text{scale and shift} \tag{9.7}$$

The proposed network also used batch normalization after every dense layer in an alternate arrangement except of the middle part which used dense layer, then highway layer, and after that, batch normalization function. The advantage of this is that it sometimes acts as a regularizer, and thus, there is no need of applying dropout as said by Ioffe and Szegedy.[8]

9.3.3 EXPERIMENTAL SETUP

This section presents an insight on how this proposed approach can take the images of the plants in real time and then classify it as diseased or nondiseased leaf.

The above architecture shows the plant monitoring setup. It can be seen that first, we will already have the trained data where the data are taken from the train directory consisting of dataset.[13] Then, the testing data have been taken in real time from a video camera and then stored in test directory. This makes the data little complex as the training and testing data differ much. The number 1 specifies that the training data have been inputted first in the CNN and number 2 specifies that after training data, testing data have been inputted in the CNN.

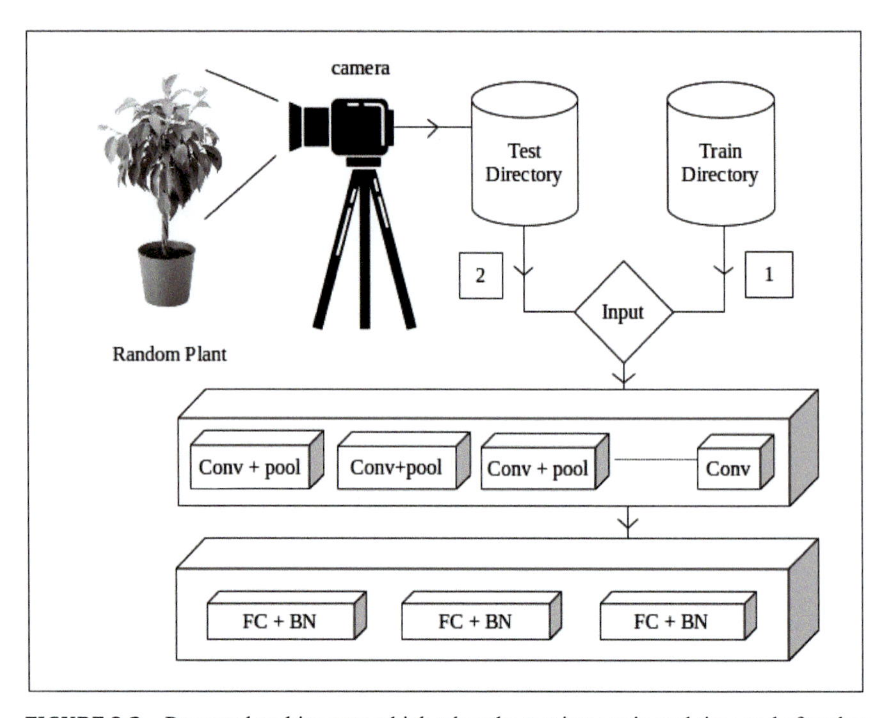

FIGURE 9.3 Proposed architecture which takes the test images in real time and after that a novel approach has been used for feature extraction and classification task.

The neural network model consists of stacking of convolutional layers with overlapping pooling layers. The number of filters used is 32 as we are dealing with small amount of pixels. The number of filters gradually increases as we move on to next convolutional layers like 64, 128, and so on. The value of stride in specified as 1 as this parameter make in the denominator part and the outcome of $[(W - F + 2P)/S] + 1$ should not be a decimal where W means the dimension of input image, F means filter size, P means padding, and S means strides. The padding is specified as valid means 0. This makes the neural network a tree-like structure and thereby decreases the time complexity of it.

Followed by this, three[4] fully connected layers consisting of 32, 64, and 128 number of neurons are used to make it also a tree-like structure, and batch normalization function is also used with it as this reduces the oscillation of the gradient when it approaches toward the global minimum and therefore reduced the problem of vanishing gradient.

Also each convolutional layer consists of CReLU activation function and fully connected layers consist of ELU activation function. In the above diagram, conv+pool represents convolutional layer plus pooling layer and 256×256 represents fully connected layer plus batch normalization function.

9.4 RESULTS

9.4.1 DATASET

First,[13] dataset has been used. It consists of different types of diseases of a plant but for our case, it has been labeled as diseased and nondiseased leaf.[14] Dataset has been used to classify different types of plants and the same approach has been implemented to get accurate performance. This is done to prove the robustness of the proposed novel approach.

9.4.1.1 DATASET 1

This dataset consists of types of diseases in different types of leaves and originally consists of 28 numbers of classes, but for the monitoring purpose, the labels have been specified as diseased and nondiseased leaf.

FIGURE 9.4 Different types of diseased leaves taken into account from the first 15 classes.

Source: Reprinted from Ref. [13]. Open access.

Non-diseased Leaves

FIGURE 9.5 Different types of nondiseased leaves taken into account from the first 15 classes.

Source: Reprinted from Ref. [13]. Open access.

Originally, the dataset consists of 28 classes which make the dataset extremely huge, but due to system configuration constraint, first 15 classes have been used. All the images have a dimension nearly around 256×256 pixels but are converted to 80×80 pixels for fast processing. Class 1 consists of diseased leaf and class 2 consists of nondiseased leaf. The overall dataset consists of 8310 training images and 3184 testing images.

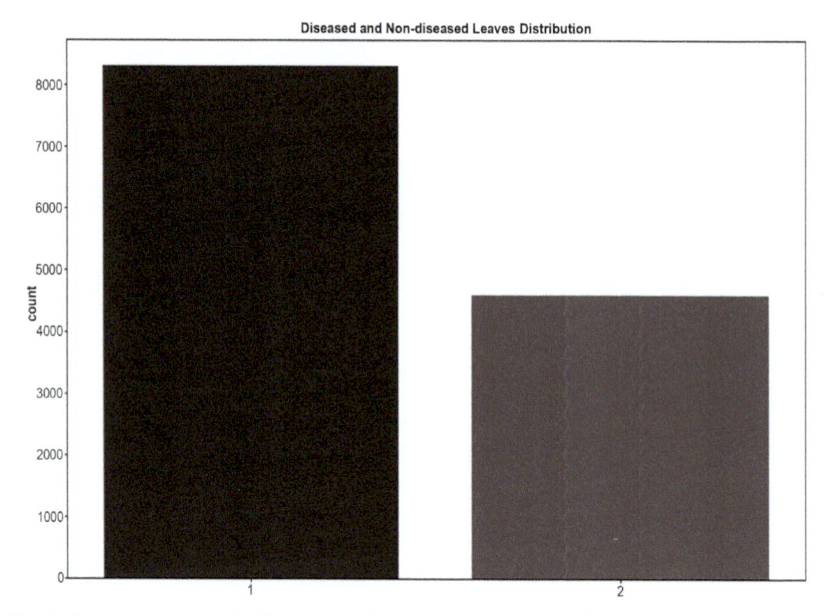

FIGURE 9.6 Dataset distribution of diseased and nondiseased classes.

The x-axis shows the numbers 1 and 2 which points toward the classes, diseased and nondiseased, and y-axis shows the count of images in both the classes. This dataset was utilized broadly to test the different features in light of the fact that measure of the dataset and the accessibility of class data takes into account execution assessment.

9.4.1.2 DATASET 2

This dataset consists of types of leaves and originally consists total of 15 classes but to check the behavior of the models in less data and different number of classes when compared to others, only five classes has been taken into account.

FIGURE 9.7 Five different types of leaves from five classes.[14]

All the images have dimensions around 256×256 pixels but are converted to 80×80 pixels for fast processing. Class 1 specifies *Acer* leaf, class 2 specifies *Alnus* leaf, class 3 specifies *Betuala* leaf, class 4 specifies *Fagus sylvatica*, and class 5 specifies *Populus* leaf. The overall dataset which is taken into consideration consists of 264 training images and 111 testing images.

The x-axis shows the numbers which point toward the classes, and y-axis shows the count of images in all five classes.

9.4.2 ESTIMATION OF MODELS BASED ON TRAIN AND TEST LOSS AND ACCURACY

Different models are used, thus each and every model are trained up to 25 iterations just to identify the behavior. After that, a final experiment was conducted where the training of the data has been done up to 70 iterations.

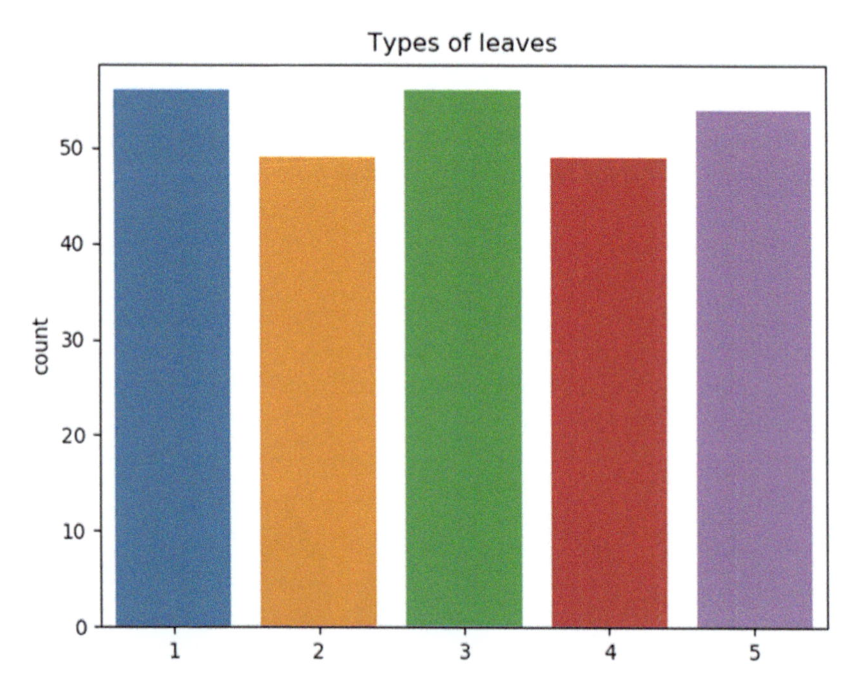

FIGURE 9.8 Data distribution of five different types of leaves.

For the optimization, Adam optimizer is used with a learning rate of 0.001.[12]

Monitoring of plants consists of identifying types of leaves and if it is diseased or not. It also consists of identifying type of disease but when it comes to the treatment domain. This chapter specifies only the monitoring part, thus only first two cases are taken into consideration.

Further experiment was also performed where the acquisition of leaves of some random plant has been performed in real time to get the test data. These data consist of 200 test images. For the training purpose, pervious dataset which is labeled as diseased and nondiseased has been used. In addition, before training, the model has been fine-tuned by dividing it into 7810 training images and 500 validation images for dataset.[13] For dataset, it has been divided into 234 training images and 30 validation images.[14]

This dataset was trained with a different number of convolutional layers. 1 Conv signifies one convolutional layer, 2 Conv signifies two convolutional layers, and so on. The experiment was stopped up to five convolutional layers as after two convolutional layers, results were not coming good. It can be seen that the four convolutional gave better result than three convolutional layers but the result at some iteration varied too much. At iteration 8, 18, and 21, the test accuracy was 0.6380, 0.6340, and 0.6960 which are less when compared to other iterations.

TABLE 9.1 The Classification Result of Diseased and Nondiseased Leaves.

Models	Train loss	Train accuracy	Test loss	Test accuracy
1 Conv	0.0663	0.9133	0.0994	0.8624
2 Conv	0.0565	0.9268	0.0929	0.8638
3 Conv	0.0571	0.9255	0.1504	0.8188
4 Conv	0.0549	0.9293	0.1149	0.8582
5 Conv	0.0562	0.9269	0.2509	0.7169

TABLE 9.2 The Classification Result of Five Different Types of Leaves.

Models	Train loss	Train accuracy	Test loss	Test accuracy
1 Conv	0.0928	0.6840	0.1242	0.5060
2 Conv	0.0866	0.7246	0.1233	0.5199
3 Conv	0.0833	0.7323	0.1357	0.4653
4 Conv	0.0833	0.7507	0.1471	0.4200

For this dataset also, the same approach has been applied to check robustness and as expected; two convolutional layers has given better performance. This experiment was stopped at four convolutional layers as gradual decrease in performance has been obtained.

The decrease in performance for three and more number of convolutional layers is due to the problem of overlapping.

TABLE 9.3 The Classification Result of the Diseased and Nondiseased Leaves Acquired in Real Time.

Model	Train loss	Train accuracy	Test loss	Test accuracy
2 Conv	0.0402	0.9490	0.0871	0.8876

The last experiment was conducted up to 70 iterations using the best model obtained from other two experiments to show that the proposed novel approach can be also performed better in data acquired in real time.

Following graphs are obtained during the final experiment of acquisition of real-time tests data:

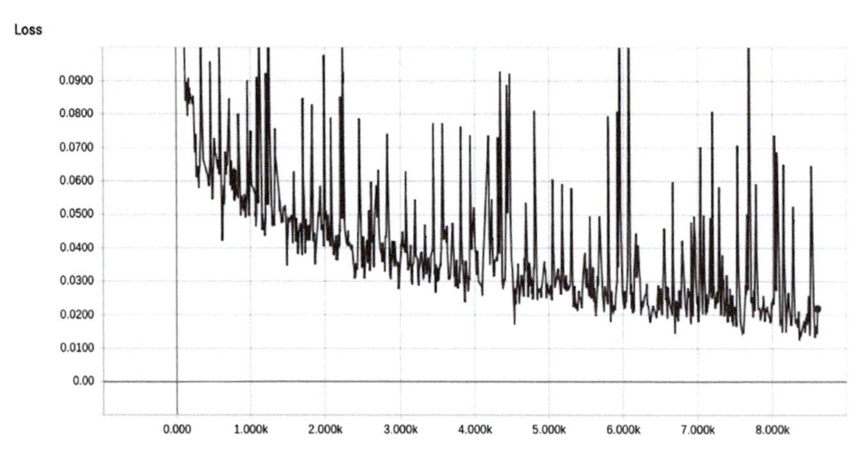

FIGURE 9.9 Training loss where the model was trained up to 70 iterations. x-Axis reflects the training steps and y-axis reflects loss values.

The results are obtained by using tensor flow's high-level API known as tflearn, a Deep Learning framework, developed by Google, and the graphs are obtained through Tensorboard except the confusion matrix.

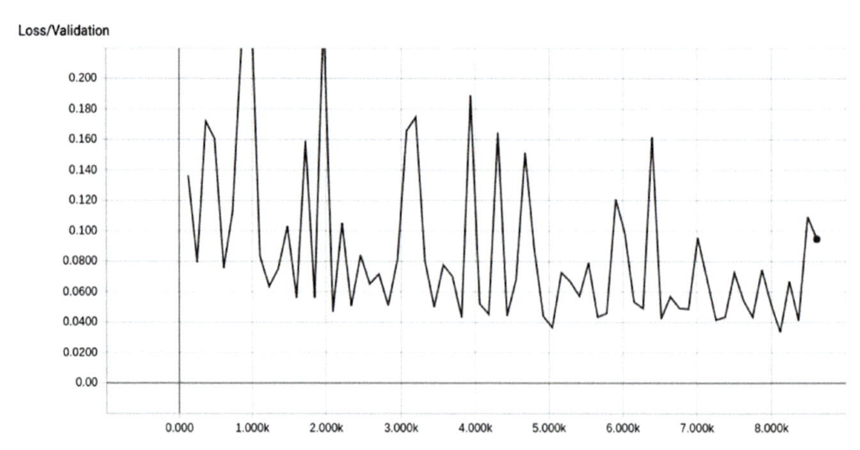

FIGURE 9.10 Graph shows the test loss where x-axis reflects training steps and y-axis reflects loss values.

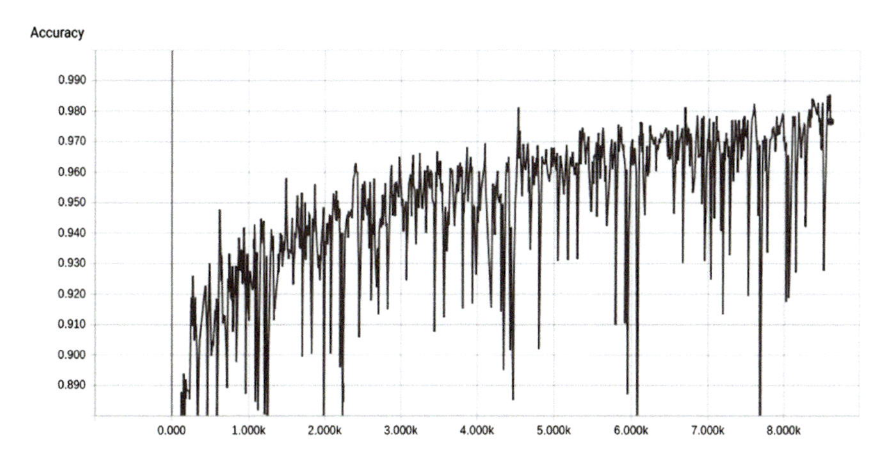

FIGURE 9.11 Graph shows the train accuracy where *x*-axis reflects training steps and *y*-axis reflects train accuracy values.

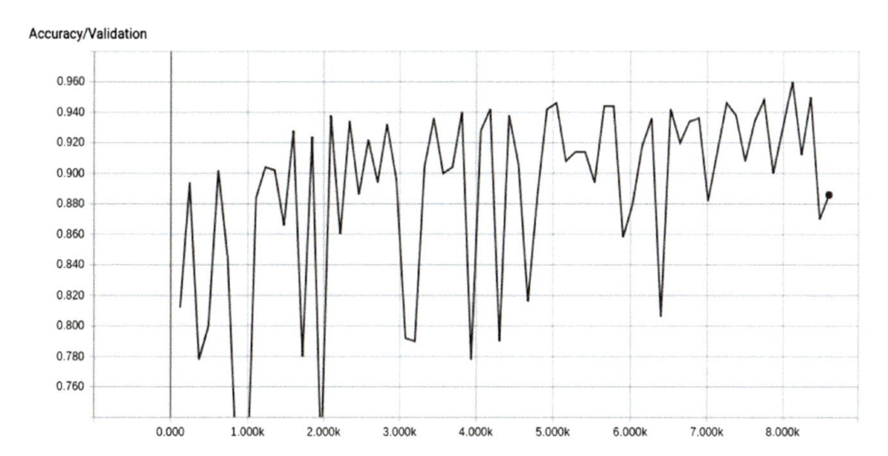

FIGURE 9.12 Graph shows the test accuracy where *x*-axis reflects training steps and *y*-axis reflects test accuracy.

9.5 CONCLUSION

From the result section, it can be seen that for dataset,[13] neural network model with two convolutional layers consisting of CReLU activation function proved to be better that the rest with the highest test accuracy of 0.8638 and least test loss of 0.0929. It was also tested on dataset[14] where model with two convolutional layers consisting of CReLU activation function gave the best result with highest test accuracy of 0.5199 and least test loss of 0.1233, thus showing robustness. Further experiment was conducted where the test data have been taken in real time and was classified by using two CNN after checking its behavior generated from the above two datasets.

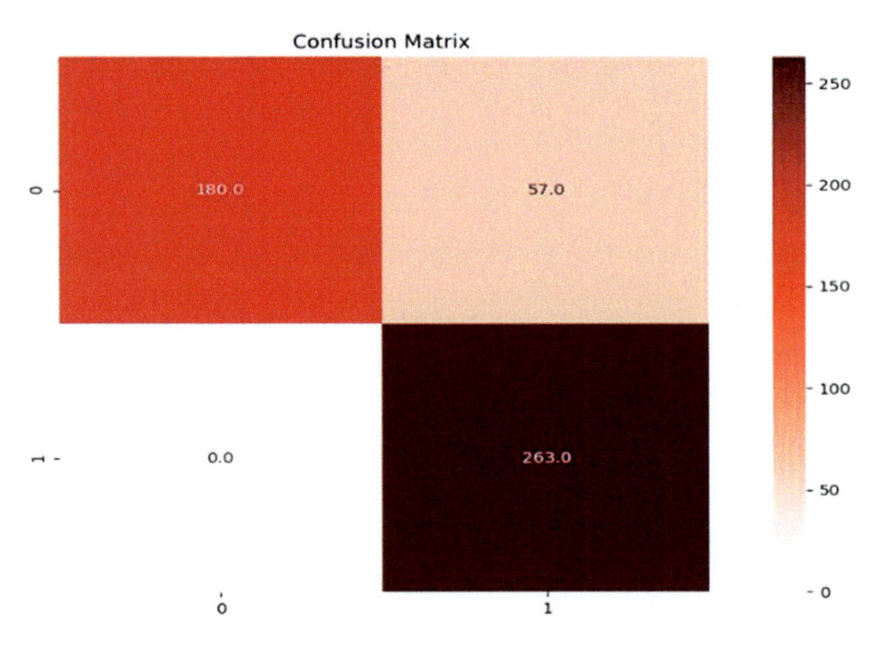

FIGURE 9.13 Heat map of confusion matrix acquired from the proposed approach when test on the data taken in real time. The *x*-axis shows the predicted labels and *y*-axis shows the actual labels. The diagonal values show the true positive values, means the chances that they are actually correctly predicted.

KEYWORDS

- **greenhouse**
- **convolutional neural network**
- **deep learning technique**
- **monitoring system**
- **leaf**

REFERENCES

1. LeCun, Y.; Bengio, Y. Convolutional Networks for Images, Speech and Time-Series. *The Handbook of Brain Theory and Neural Networks*, 1998.
2. Gu, J.; Wang, Z.; Kuen, J.; Ma, L.; Shahroudy, A.; Shuai, B.; Liu, T.; Wang, X.; Wang, G.; Cai, J.; Chen, T. Recent Advances in Convolutional Neural Networks. *Pattern Recogn.* **2018**.
3. Ramchoun, H.; Idrissi, M. A. J.; Ghanou, Y.; Ettaouil, M. Multilayer Perceptron: Architecture Optimization and Training. *Int. J. Interact. Multimed. Artif. Intell.* **2016**.
4. Kakde, A.; Arora, N.; Sharma, D. Novel Approach towards Optimal Classification Using Multilayer Perceptron. *Int. J. Res. Eng., IT Soc. Sci.* **2018**.
5. Clevert, D.-A.; Unterthiner, T.; Hochroiter, S. *Fast and Accurate Deep Network Learning by Exponential Linear Unit (ELUs)*. In Proceedings of the International Conference on Learning Representations (ICLR), 2016.
6. Pedamonti, D. Comparison of Non-linear Activation Functions for Deep Neural Networks on MNIST Classification task. **2018**. arXiv:1804.02763v1 [cs.LG].
7. Shang, W.; Sohn, K.; Almedia, D.; Lee, H. *Understanding and Improving Convolutional Neural Networks via Concatenated Rectified Linear Units*. In Proceedings of the International Conference on Machine Learning, 2016.
8. Ioffe, S.; Szegedy, C. Batch Normalization: Accelerating Deep Network Training by Reducing Internal Covariate Shift. **2015**. arXiv:1502.03167v3.
9. Kamilaris, A.; Prenafeta-Boldú, F. X. Deep Learning in Agriculture: A Survey. *Comput. Electronics Agric.* **2018**.
10. Rodríguez, S.; Gualotuña, T.; Grilo, C. A System for the Monitoring and Predicting of Data in Precision Agriculture in a Rose Greenhouse Based on Wireless Sensor Networks. *Proc. Comput. Sci.* **2017,** *121*, 306–313. DOI:10.1016/j.procs.2017.11.042.
11. Geng, L.; Dong, T. An Agricultural Monitoring System Based on Wireless Sensor and Depth Learning Algo. *Int. J. Online Eng.* **2017,** *13*, 127–137.
12. Krizhevsky, A.; Sutskever, I.; Hinton, G. E. ImageNet Classification with Deep Convolutional Neural Networks. *Neural Inf. Proc. Syst.* **2012**.
13. Hughes, D. P.; Salathe, M. An Open Access Repository of Images on Plant Health to Enable the Development of Mobile Disease Diagnostics. *Comput. Soc.* **2016,** arXiv:1511.08060 [cs.CY].
14. Söderkvist, O. J. O. *Computer Vision Classifcation of Leaves from Swedish Trees*. Master's Thesis, Linkoping University, 2001.
15. https://www.funnydog.tv/video/green-house-monitoring-and-controlling-systen-using-gsm/U7xphzYiD-c.
16. http://www.pestnet.org/SummariesofMessages/Crops/Rootstubers/Taro/Fungi/Taroleafblight,Samoa.aspx.
17. Stewart, M., Simple Introduction to Convolutional Neural Networks. Toward Data Science, Feb 26, 2019. https://towardsdatascience.com/simple-introduction-to-convolutional-neural-networks-cdf8d3077bac

CHAPTER 10

Application of Artificial Intelligence in Image Processing

AAFREEN NAWRESH* and S. SASIKALA

Department of Computer Science, University of Madras, Chennai, India

**Corresponding author. E-mail: anawresh@gmail.com*

ABSTRACT

Artificial intelligence (AI) can be portrayed as the one which makes computer programs to be able to think, behave, and also perform certain tasks that only humans can do. AI in imaging technology is helpful to perform analysis such as recognizing the feature of object like shapes, size, textures, and so on just to understand what it really is. AI has made wonders in the imaging field by providing quick diagnostic analysis, depiction of the upcoming events, financial graphs, facial recognition, robust gaming compatibility, amazing virtual reality experience, effects in entertainment, and much more. The existing AI-based imaging technology has seen to be slowly improving the digital world experience. As technology improvises the other fields, such as neural network, computer vision, machine learning, big data seem to merge with AI to bring out the best from every field. Though AI was introduced during the 20th century, it is slowly attaining its popularity only now, due to the beginning of the big data. As working on huge data needs larger compatible systems with better processing speed and storage mediums, AI helps in achieving this. Imaging technology using AI has steadily made researchers believe that there will be growing developments to make a better world.

10.1 INTRODUCTION

Artificial intelligence (AI) was coined by John McCarthy during the 20th century in a conference at Dartmouth, where he stated it as "Every aspect of learning or any other feature of intelligence can in principle be so precisely described that a machine can be made to simulate it. An attempt will be made to find how to make machines use language, form abstractions, and concepts, solve kinds of problems now reserved for humans, and improve themselves." In general, an AI system is a machine that will achieve jobs which a human can perform with their intellect. The task of achieving or accomplishing targets relies on the training given to the systems which learn to follow certain rules or protocols when data is given. It also involves targeting self-realization of errors and correcting it to attain proper and accurate results. AI is an emerging field which invokes researches for human reasoning in AI systems. Researchers have tried to create major pace in the development of successful AI applications and systems. One must know that the advanced systems of AI came from neural networks which were developed in the 1950s. The reason why it is a major hit these days is because of the beginning of the big data, since many machine learning and neural network systems need heavy processing speed and storage medium, which can be achieved through big data efficiently. Precisely, storage and computing costs get reduced effectively. In the recent era, AI has entered into various fields from text recognition in images, language processing to the disease prediction algorithms using image processing. In the field of imaging technology, simple road traffic sign board classification, processing of vehicle number plate captured on CCTV footage, identifying a particular person using sketch and also via CCTV footage. In general terms, AI mastered itself through machine learning, deep learning, natural language/linguistic processing, robotics, software engineering, and many more.

As of now the target of concentration is on image processing, just as the saying goes, "Rome was not built in a day," the present generation imaging technology also has a history that started back then. The first camera was the Camera Obsura, which used pinhole or lens to develop images in an inverted way on the surfaces. The Obsura camera developed in the year 1544 was first used by Reiners Gemma Frisius to observe the solar eclipse. Later on, it was suggested to be used as a drawing aid by Giovanni Batista Della Porta in the year 1558. Then came the time where one can only take pictures through camera and not preserve them for long. The huge sized camera was identified by Johann Zahn in the year 1685. The first-ever camera which not only captures but preserves the pictures taken was invented by Alexandar Wolcott in the year 1840. The sliding window wooden box camera came to the world, which was created by Joseph Nicephore Niepe, who used a coating of silver chloride on the paper that darkened when exposed to light. This technique failed to get established as it was not a solution to preserve photos. Thus, to enhance the existing sliding wooden box, a new wooden box was created by Charles and Vincent in the year 1826.

Nicephore was keen enough to bring about change in the photography era. In 1836, Nicephore along with Loius Daguerre created a working sensible photography method known as Daguerreotype, which was coated silver on a plate of copper, which was then indulged with the iodine vapor such that it can be sensitive to the light. The image was finely developed using a salt solution along with mercury vapor. Then came an attempt to improve the working known as Calotype, developed by Henry Fox Talbot, during the year 1840. To create handy cameras, Richard Leach Maddox competed with the collodion dry plate and gelatine dry plate model and surpassed the working with the wet plates. The exposure time was also lowered so that perfect photos were created. George Eastman (Kodak Firm) first launched the paper film during the year 1885 and was later changed to celluloid in the year 1889. Finally, in the year 1888, camera developed by Kodak came into the selling market. It had the features of fixed focusing, without flash but at a low price, which attracted the consumers. In the year 1952, the first-ever Japanese SLR with 35-mm film named Asahiflex was launched. Very soon other Japanese companies producing cameras namely Canon, Nikon, Yashika also took a step into the market. Nikon was also being ranked as the professional quality equipment company.

With the advancement in technology for the creation of imaging tool, the Polaroid Camera model 95 was introduced in the year 1948. Polaroid (the instant camera) invented by Edwin Land used a chemical process that helps in printing the picture taken from the exposed negative within a time of about 60 s. During 1980s, digital cameras like Kodak DCS-100, Fuji DS-1P, and Dycam came to the market which had liquid crystal display (LCD) screens with the format of JPEG for image and MPEG for video recording, were used, and had storage capacity on memory cards with storage space of about GBs to TBs rather than using films. In this era, smartphone cameras with a capturing resolution of about 20 MP are used. At present, Canon developed a digital camera with an incredible 120 MP resolution capacity, where every minute detail can also be captured, also a new video format with about 16 times the resolution of current HD televisions.

The evolution of television luckily started from the advent of camera and its video-recording facility; the British High Definition Television service started providing provision to watch television (black and white) by 1936. Since then, the development in the world of Television has been promisingly enhanced, from black and white (using mechanical and electronic system), color television (1953), LCD (used cold cathode fluorescent lights in a super-thin display technology screen), LED (light-emitting diode, using the light-emitting diodes to display with less heat produced), HD TV (high definition, which promises to provide crystal clear clarity and amazing surround sound, the resolution provided ranges from 720 to 1080 progressive scan per frame, finally we have ended up in making "The Idiot Box" our favorite time pass).

In the past century, medical imaging techniques have been transforming the medical field with technologies that made doctors believe that there is an assurance that their prediction is true. In the field of Medical Imaging Technology, X-rays, computed tomography (CT), and magnetic resonance imaging (MRI) scans were introduced where each one is the advancement of another in order of development. Probably, with the fast growing phases of imaging innovation, one needs advance methods which will help in portraying the real affected/infected area. X-Rays (X-radiation) are the most general and broadly available technique used for diagnosis. Even though there are many refined tests, X-ray is the most affordable test that is taken first as a primary diagnostic procedure. The expansion started with CT, which combines X-rays and computer technology to give a more precise view of cross-sectional region of body. A CT scan helps the doctor to see the affected region's shape, size, and the position that has affected the body like tumors, tissues, hemorrhage. A CT scan generally costs a little more than the existing X-ray and also takes time in producing results. On the contrary, to get a clear and deep image of the tissues and nerves, a full-body MRI machine was invented, which was named "Indomitable," which means impossible to defeat. It sure by name is impossibly the best imaging system since then.

The start of pinhole camera made various advances and we are here where good resolution cameras come handy anytime and can also capture details using drones through a remote operation. A lot of development seemed to happen by then and even everyday people believe the better is yet to come. For every start, there is a new beginning and end to the old, the end has not occurred still and things seem to change for the better as AI has entered into all of the dimensions.

10.2 EXISTING TECHNOLOGIES AND ITS REVIEW

AI arises out as a modifiable technology of our digital era. Doubts on how it works, what it does and what criteria does it have that makes it an essential technology across business, education, medical, science, psychology, imaging, cinematography, robotics, and much more. AI seems to be of great interest to almost everyone, who wishes to gain in-depth knowledge on it. There are enormous amounts of articles, discussions, headlines every now and then about the benefits and effects of the technical advances it has made. Even still people feel that AI may rule over the world, it seems to keep moving forward, bringing up applications to smartphone from checking out the weather, facial recognition, AI face unlocking, detecting hypertension, and diabetes, and so on. Much importance is given to science-related AI, such as the robots performing surgery, the other unnoticed applications of AI that are already in use by an individual goes unaware; let us know about them now.

At the time of World War II, renowned British Computer Scientist Alan Turing worked on to decode the code "Enigma" which was encoded by German forces to pass messages securely.[1] Alan Turing along with his team developed the Bombe Machine which was helpful to decipher Enigma's messages. It was the Enigma and Bombe Machines that laid the foundation for machine learning. According to Alan Turing, a machine that has the capacity to converse with humans without the humans knowing that machine would win the "imitation game" and could be said to be "intelligent." As already said, John McCarthy was the one responsible for coining the term "Artificial Intelligence," the research centers aroused across United States to discover the prospective of AI. Following which researchers Allen Newell and Herbert Simon gave an active participation to promote AI as a field of computer science which would transform the entire world, they developed a General Problem Solver algorithm which can solve mathematical problems.

The rise of research in AI started in the 1950s and 1980s, where computers were given instructions manually on recognizing images, objects in the images, features needed.[2] This system of traditional algorithms is known as Expert Systems, as mainly they require humans to identify features for every exclusive scene of object that has to be used to represent the feature in mathematical models that a computer can understand. It involves a lot of hard work since there are enormous ways an object can be represented and also many different scenes and objects that are unique, and finding the accurate model to characterize all the possible features of every object or scene, and for every possible detected object or scene is more of a work that may remain forever. In the late 1990s, machine learning was introduced where instead of programming computers for what has to be checked in identifying the scenes and the objects in videos and images, one can model an algorithm that will formulate computers to learn how to recognize object in an image, a scene by self, just as a little child learns to know about the environment by investigating things. Machine learning unlocked the approach for computers to learn to identify approximately any scene or object that is required.

In the year 1986, a paper entitled "Learning Representations by Back-propagating Errors," Rumelhart, Hinton, and Williams described the detailed process and working of backpropagation.[3] The paper showed how it could immensely progress the existing neural networks in many jobs like the word prediction, shape recognition, and many more. In spite of some drawbacks after the early success, Hinton kept on with his research at the Second Artificial System Winter meet to attain new altitude of achievement and compliments. He is considered by a lot of individual as the "Godfather of deep learning."

The most popularly used algorithm support vector machines (SVMs) has been making rounds since 1960s, it was used and also modified by many.[3] The current standardized model was designed by Cortes and Vapnik in 1993 and represented in 1995. The SVM is a system used in recognizing and mapping similar data and is used for handwritten character recognition, text categorization, and classification. It has been used over many decades in all of the research findings, and many have quoted that SVM is the only algorithm that works well and gives optimistic results for any type of dataset.

Kunihiko Fukushima is known for the creation of Neocognitron, an artificial neural network, which learned to recognize visual patterns.[3] It was assigned to be used for handwritten character and other pattern recognition tasks, natural language processing, and also recommender systems. The work carried out inclined many toward neural networks, and this helped Hubel and Wiesel to develop the very first convolutional neural network, which was based on the visual cortex organization that is found in animals.

Yann LeCun, another important scientist in AI and Deep Learning World, combined the recent backpropagation method and convolutional neural network to be able to read handwritten digits, during the year 1989. The method proposed was used to read handwritten cheques and zip codes; it was helpful to process around 10–20% of cheques in the United States during the 1990s and 2000s.[3]

The GRAVA architecture developed by Robertson was created to evaluate interpretation of problems.[4] An interpretation problem is the one where the given input, for example, will be an image or a signal, which must be inferred in the perspective of domain information to construct a representative symbol. The GRAVA architecture has formerly applied effectively to the understanding of satellite aerial images and is custom-made to provide believable results to the interpretation problem of assembling a reading of the Vindolanda texts. Since then, people have tried to get to know about old scriptures, letters, ink and stylus texts, and inscribed texts. And the interesting thing is they have made a dictionary for all the identified alphabets and letters, so whenever an archeologist finds out an ancient stone or even a piece of paper with text written, they can make sure those patterns match with their existing texts from the created dictionary.

Computers are not trained to cure or treat Cancer, but they play an important role in helping diagnose it. The computer-aided diagnosis prototype, an intelligent model, was developed at the University of Chicago, where it took over 22,000 mammogram images and detected about 52% of cancer images accurately, more than the radiologists identified them.[5]

Fei-Fei Li, an eminent professor and head of the AI lab at Stanford University, launched ImageNet in the year 2009.[6] ImageNet is basically a database containing labeled images. By the end of 2017, the ImageNet consists of more than 14 million labeled images which were made available to students, educators, and researchers. Labeled data are in general those that are to be trained using the supervised learning process. Images were labeled and organized using the WordNet, which is a lexical database of English words where verbs, adverbs, nouns, and adjectives are sorted and grouped by synonyms called synsets.

The famous "Cat Experiment" that happened in the year 2012 by Google's X lab, was a key step forward, in the world of neural networks.[7] Where, using deep learning algorithm on neural network on over 16,000 computers, the team took over 10 million random images/thumbnails from YouTube which were unlabeled and gave as input to systems and made it to run, the network began to search for cats. When the session got over, the unsupervised learning had made the system to learn and train itself to identify and recognize the cat images. The network acquired accuracy of about 81.7% in detecting faces of humans, 76.7% accuracy in detecting human body parts, and about 74.8% accuracy in identifying cats.

In the year, 2014, DeepFace, a face recognition feature, was developed by the most widely used social networking site: Facebook.[8] It used neural networks to identify faces with an accuracy of about 97.35%. With the existing image recognitions modality, Facebook has improved the system by 27%, coming to 97.50%. First, a template will be created by going through the images of an individual, which includes the users' profile photo and images that are tagged by friends. On uploading a new photo, similarities or patterns are being matched with the existing template and it automatically tags friends if the template matches. This feature also enables one to unlock an account where the user has to identify the photos of friends when their account has been hacked. Seeing this technology grow, Google Photos have also adapted this type of program.

A team from the Department of Computer Science, University of Oxford, formulated an AI system and named it as LipNet.[9] The system was built on the data set known as GRID, which was made up of face forward videos of 3 s of people reading sentences. Each sentence was based on a string of words which follow the same pattern. The team used the 3-s clips to train the neural network, alike the method used in speech recognition. The neural network recognizes differences in shape of the mouth, learning to link the information to provide an explanation of the speech made in the clip. AI does not analyze the clips in a grasp but it considers the whole dataset to enable it to attain an understanding of the circumstance with the help of the sentence that is being analyzed. It is an important task since there are few mouth shapes than the noise produced by the human. When testing the system, it identified about 93.4% of the words accurately, while the human lip-reading experts performed the same task and they identified the words with only 52.3% accuracy. Through this approach, the department from Oxford went on to work with Google DeepMind, it used a series of 100,000 videos taken from BBC Television. The videos attained a clearer and broader approach of language, since they had differences in lighting and positions. The AI managed to identify words with about 46.8% accuracy, whereas the human experts provided an accuracy of about 12.4% only. AI clearly proved to perform well in lip-reading than humans' potential.

Google Maps were initially dependant on data received from traffic sensors, which were mostly installed by Government transport departments or private companies that targeted in accumulating traffic data.[10] The sensors were able to detect the size and the speed of the vehicles passing by through the radar, active infrared or laser radar technology to further wirelessly transmit that information to a server. Data from these sensors were used to provide real-time traffic updates and also the collected information becomes part of the data source for historical data's that may be used to predict the traffic update on the upcoming dates. Sensor data were limited to primary roads and highways since the sensors were only installed in the most busy roads or traffic prone zones.

In the recent approach, there have been models that learn and create data by themselves, given an input data, similar type of data are being created, such type of models are known as the generative models.[11] But training models to create or make data in this way are not that easy, but in this developmental era, a number of processes have initiated to work quite fine. One advanced approach is by using generative adversarial networks (GANs). The AI director of Facebook, Yann LeCun recently said that GANs are the most important development in deep learning. It was introduced in the year 2014, by a team of people from the University of Montreal; the main scope of GAN was to have two neural network models competing with each other. One model works in taking up image as the input and generates samples (generator), while the other model will receive samples from the generator and the training data, and finally be able to differentiate between the two sources. The two networks operate in a continuous manner, where the generator learns to produce many samples, and the discriminator learns to give good and distinguished generated data from the original data. As of now, GANs have been initially applied to model natural images. The results produce are effectively good than those trained using generative method using the maximum likelihood training methods.

Over the past 15 years, companies like Amazon, Google, Baidu, and others started using machine learning for their huge commercial benefits.

These companies not only focused on understanding consumer behavior, their hits and clicks but also targeted on computer vision, natural language processing, and many other AI applications. Machine learning is now being incorporated in many of the e-commerce websites that are used. Many initiatives have been emerged which promises to provide easy and stress-free lifestyle, like online food ordering systems, where we can visually see the type of food we want, online clothing suggestion system, where we can try outfits to a model which has our similar features, online eyewear selection system, where we can upload our image and select a suitable eyewear from many varieties, and many more. These companies are involved not only in processing user data to understand consumer behavior but also have continued to work on computer vision, natural language processing, and a whole host of other AI applications. Machine learning is now embedded in many of the online services we use. As a result, today, the technology sector drives the American stock market.

10.3 APPLICATIONS OF AI

AI has stepped into the human world, making changes in common life-style. Every AI scope has a more benefitting future attached to it; the development occurs when people get more interested in the working of it and create new models out of it.

10.3.1 HEALTH CARE

One such system is the medical body scanners, where after X-ray, CT scan, MRI scan, deep-learning startup organization **Aidoc** has created the world's first and only complete, full-body scanning solution using AI which helps to analyze CT scans, for highlighting the findings to radiologists. It offers support to radiologists for scanning areas such as the head, C-spine, chest, and abdomen.

AI can now be used to effectively save your life. AI is being used to scan and read a patient's medical data and help in predicting if that patient is vulnerable to heart attacks and strokes. A recent study has even found that AI seems to be more accurate than the doctors while making such kinds of predictions. The AI looks through the patient's medical records and checks like a cross-reference with existing historical data of other patients who have already being affected from heart attacks and strokes. The results showed that AI made 355 more correct predictions than those made by doctors. This seemed to be surprising since the processing capabilities of computers were compared to the human brain. In effect, it could also predict one day that you might die.

DeepMind has bought many healthcare projects across the world; now in collaboration with UCL's radiotherapy department, it has initialized to reduce the amount of time taken to plan treatments.[12] Through machine learning, DeepMind has given access to 1 million images of the eye scans, along with their patient data. It sets to train itself to read the scans and predict spot early signs which may indicate the occurrence of degenerative eye and also reduce the time taken for diagnosis is reduced to one-fourth.

Entrepreneur Jonathan Rothberg of **Butterfly Network** proposed to create a new handheld medical-imaging device that aims to take both MRI and ultrasounds scans easier and in cheaper rate.[13] Butterfly iQ device uses semiconductor chips, instead of piezoelectric crystals. It will use ultrasound scanners to create 3D images, then sends the images to a cloud service, which will further work on enhancement, zooming in on identifying features in the images and help to automate the diagnoses. The service will incorporate deep learning techniques into its device. The ultrasound-on-chip technology will replace the existing transducer and the system with a silicon chip. The plan is to even automate many of the medical imaging procedures, the device will be made available for usage in clinics, retail pharmacies, and in poorer regions of the world with an affordable pricing of about $1999.

An application programming interfere (API) Clarifai powered with AI provides scene recognition in videos.[14] In images, it does text recognition, logo detection, face detection and sentiment analysis, as well as a more powerful version of image detection such as color, brightness, and dominant color. The existing model only identifies the predominant faces, but the new model detects all the faces that were captured far away from camera, taken at different angles and also partly seen. This face detection model with the prediction API in return give bounding box coordinate regions of human faces that were detected in the input image data. The API is made from a simple idea. When a prediction is made through API, one has to suggest the model that has to be used. Generally, the model consists of concepts. A model will only "check or see" the concepts that it consists of. In addition, one can specify more variables for predictions. An input (image or video) is sent to the API and it in turn provides the predictions. The prediction is based on the model one runs the input through. For example, if the input is run through a "car" model, the prediction will definitely return the items/features the "car" model knows. For a video input, the API response provides a list of predicted concepts for each frame of the video. The processing of videos occurs at 1 frame/second, which means that for every second of the input video, a list of concepts will be returned. Even here the prediction on the video is done using the available models. The models that are supported currently are general, apparel, food, wedding, and travel. At any time, one can create an API call just by initializing the {model-id} variable and the input data variable is video here. The prediction for video is limited to length and size it supports. A video, uploaded through any other link or URL can contain size up to 800 MB or length of 10 min. The most common criteria for using Clarifai are to initially get the concepts predicted in an image and then use those in searching. The search API allows one to send images to the prediction service and make them indexed in "general" model concepts and also its visual representation. Once the data gets indexed, one can search for image by concepts or by the image itself. When the images with concepts are added, one can start creating the model.

Apple Watch users can now take a look at their hearts rhythm just by holding the crown of the device.[15] The software update provided to the Apple Watch Series 4 provides a new feature, to identify the atrial fibrillation, also provide extrapassive monitoring. People over the age of 22 and above can utilize the features provided to differentiate between a normal heart rate or with atrial fibrillation and sinus rhythm. An optical heart sensor uses green LED lights paired with light-sensitive photodiodes helpful to detect blood-volume pulses in the wrist of a human using the "photoplethysmography"-based algorithm, which is an easy and inexpensive optical method that is used to detect blood-volume variation in the microvascular area of tissue. To verify heart rate variability, Apple Watch captures a tachogram, a plot of the time between heartbeats, every 2–4 h. It also allows the user to send messages to friends showing their level of heart rate in case of emergency and also recommends the user to visit a Doctor to seek proper medical care if irregular symptoms are seen.

An ultrasonography exam takes quite a lot of time in identifying the planes in the brain, which needs an ample amount of training and manual work. There could also be a missed or delayed diagnosis. Now, with AI systems, users will just need to find a starting point in the fetal brain and the device will automatically take measurements after identifying the standard planes of the brain. The data or the documentation are maintained as the patient may visit for examination some other day; this will help in more positive diagnosis.

EchoNous has developed a convolutional neural network for the automatic detection of the urinary bladder with the help of high-quality ultrasound images captured with Uscan, using the advantage of the "high spatial density fanning technique."[16] With the help of the captured image, one can compute the urinary bladder volume with much higher accuracy.

Cancer can be diagnosed promptly with the help of deep learning and AI concepts.[17] A Chinese start-up named **"Infervision,"** will use image recognition technology and the deep learning to efficiently diagnose the signs of lung cancer with X-rays as the same as Facebook recognizes faces in photographs. Infervision products are claimed to be designed using the advanced deep learning algorithms, convolutional neural network, to imitate the human cognitive performance. The images after getting trained with the data of great high quality and measure, the model will routinely detect the patterns with a high-speed estimation potential, and it develops into more intellectual through iterations. A doctor can identify a number of diseases through an image scan; AI has to get trained on how to identify numerous target objects in a jiffy. The attempt to build medical devices as more reliable, accurate, and automated is producing a growing interest in finding ways to incorporate AI systems. Medical imaging is an area that is progressively developing and improving devices to support the management and treatment of chronic diseases and will most likely continue to be a major area of focus.

10.3.2 GOOGLE SERVICES

10.3.2.1 GOOGLE MAPS

Google Maps have formulated a better way for helping get to work, school, or areas we wish to visit.[18] Google Maps clearly indicate traffic-free, slow-moving, or highly congested traffic for the route one is traveling. It shows the green, yellow, and red lines on the routes that one is to estimate the fastest path to reach destination. How does Google do it? Google knows the traffic conditions on the path we are and the path we are supposed to go. Google Maps gives us the information based on the existing traffic views and the criteria to take up a different or a faster route is based on two informations, one being the historical data which tells the average time taken to travel a particular part of the road at the specific time on each and every day, and the other is the real-time data that is sent by the sensors and the Global Positioning System enriched in Smartphone that provide information on how fast the cars are moving. From the start of 2009, Google started to improve the accuracy in the traffic prediction of its services. As soon as Android phone users switch on their GPS, the Goggle Maps app navigates to the user's current location and the phone sends bits of data to Google, which lets the organization, know the cars movement. Google Maps constantly merges the data that is coming in from all the automobiles present on the road (only through GPS enabled phones) and immediately sends back by way of those colored lines on the traffic layers. The possibility of getting accurate predictions relies on more usage of the app, since the traffic predictions are more likely made through the vehicles speed that is traveling in the same route, without stopping at any arena or a coffee shop. A fact that one has to know is that, if Google Maps do not have enough data to provide about a path of the road, then that path will appear as gray on map. Google Maps partnered with Waze in the year 2013 has added a new feature where the users/drivers can report about the nearby traffic accidents, slowdowns, speed traps, or even disabled vehicles. These reports appear on Google Maps as small icons showing signs like the construction sign, speed cameras, or even crashed cars. The AI incorporated Google Maps and Google Street-View sense traffic and flow of the vehicle to make riders achieve a good ride.

10.3.2.2 GOOGLE STREET-VIEW

Google also provides Google Street-View service where a typical view of the street is shown on the smartphone. Many applications use augmented reality to provide a high level of digital information of the world around us. Keeping such apps on the smartphone, one can hold the camera up to click the image of a street and further receive the data about it and its surroundings. The street view enabled cities approach comes from the company called "Immersive Media," where like interactive 3D pictures, videos also use the same technology. Eleven lens camera known as the Dodeca 2360 is used, which captures a massively huge surface area of images and at a very high resolution. The camera is generally placed on top of a vehicle which is moving and records both video and also the geographical data. What we visualize is the image stills from the captured video, which is why a new image is being seen every time we check the recorded route. As a first step, one needs to roam around and capture images of the locations to show in the street view, paying attention to details such as the weather, population of the area, and also the best region to capture the images. Further to match every image to its geographic location on the map, signals are combined from sensors through the car's GPS that measure direction and speed. This helps in locating the car's exact route and also to align the images whenever needed. To avoid the gaps in the 360° photos, cameras adjacent capture the overlapping images, and further combine the images to form a single 360° image. The images are smoothened to lessen the "seams" using image-processing techniques. When one travels to an area in the distance, the 3D model will indicate the correct panorama image to show the exact location.

10.3.2.3 GOOGLE GLASS

Google is one such company which has started creating solutions to problems in the form of a wearable device.[19] It simply looks like a spectacle with one frame being thicker than the other. Google Glass opens one's eyes to the digital world. During the year 2012, an account named "Project Glass" flashed on the social networking website—Google Plus. The account's post publicized the need of the project—in creation of the wearable computer that will help in exploring the world. The post also conveyed the concepts that the Google Glass will do. The Google Glass consists of thick area of frame in the right eye side which is the side where the screen has been inserted. To see on the screen, one has to glance up with eyes. The Google cofounders wore the Google Glasses at the Google I/O event in San Francisco on June 27, 2017, the audience were thrilled with the demonstration of technology. Google Glass owners can use the glass to perform the following operations: click and share the photos and videos, perform Google search, provide directions for every turn, get alerts from social networking websites and even text messages, get updates on the weather and traffic of that region, start a video call, get alerts of various appointments and other calendar events. As a future enhancement, the Google Glass will help to track people's happening and also the people we meet. The facial recognition software and social networking sites will possibly take a look at a person who you would have met and then see the same person's public profile on the social platforms. Google Glass is assembled with chips, feedback devices, and sensors. Google Glass can be controlled with the capacitive touchpad which is on the right side of the glass. When finger makes contact to the panel, a controller chip will detect the change in the electric capacitance and keeps it as a touch. Swiping fingers in horizontal direction will allow one to navigate the menus available in the devices, whereas swiping downward on the touchpad will get back from the choice.

The way of controlling the Glasses is via voice commands. A microphone attached onto the glasses receives the voice and then the microprocessor understands the commands. One cannot expect to get a proper respond just by saying anything, there is a list of commands that one can use, and all of which starts with the phrase "OK, Glass," which alerts the glasses that a command might soon follow. For example, "OK, Glass, take a picture" will send an order to the microprocessor to click a photo of whatever the wearer is looking at.

The processor used in the Google Glass is originally from Texas Instruments. It is called the Open Multimedia Applications Platform chip. These chips categorized from a larger categorization of microchips called the systems on chip, which means there are multiple mechanisms that are working together—here, an ARM-based microprocessor, video processors, and a memory interface. The chip is capable of playing video up to 1080p resolution and 30 frames/second. The circuit board also contains a SanDisk flash drive of memory—16 GB of storage space, but only 12 GB are made available to the user. The company known as Micron Memory (Elpida) provided the dynamic random access memory chip. These chips not only provide storage for media and apps but also the memory needed to run the programs on the glass. A single chip inside the Google Glass will provide the support for internet connection, whereas another chip called the SirFstarIV will support the GPS for determining the location through satellite signals. The camera can click photos with a resolution of about 5 MP. The video capturing has a capacity of 720p resolution. The speaker on the Google Glass is known as the bone conduction speaker, which helps the speaker to send vibrations that pass through the skull and then to the inner ear. In this case, there is no need to wear headphones or earbuds. Using this facility together one can make conference calls.

The Google Glass has also entered the Operation theatre, where doctors can see the medical information without turning away from the patients.

Few of the researchers have found that the navigating options will help find tumors by the surgeons but can also project a form a tunnel vision or blindness on a part that could make them miss unrelated lesions or the problems around them. All of these features are provided just with a charging of about 45 min where the battery has a lithium polymer battery with a capacity of 2.1 W h.

Google Glass is specially fit for providing surgical education and telementoring. Preliminary testing has proved that POV recordings are better than the standard video recordings in representing anatomy and surgical procedures. Unlike other POV cameras planned for use in sports (GoPro camera), Google Glass does not deform with a fisheye lens and can be monitored using voice and movement of head. Videos can also be live streamed for intra-operative discussions, telementoring, or secluded management of bedside procedures. Google Glasses are used in the simulation lab, due to patient privacy issues, but have complete several one-off live streams with individual patient permissions. As of now, it has been incorporated to use for training courses and providing medical education.

Future use of Google Glass may contain video recording of the surgery as part of the documentation for the history of patients medical record, or as a means for assessment of surgical competency as part of ongoing professional development and certification. It will also be helpful for giving assessment and feedback to improve surgical methods. The AI features incorporated here include image-recognition capabilities. Although we identify that Google Glass is a lacking and new expertise, the start of Glass suggests a technical change in the way we understand the world. This will make performing surgery easier in the future, and let's wait to see that happen.

10.3.2.4 GOOGLE LENS

The Google Lens technology works on the smartphone's camera to "observe" what's around you and provide you information about the area around you or your surroundings.[20] The camera on your smartphone can simply provide you contact information (capture the image of the business card and Google Lens will update the info into your mobile phone by just capturing the phone number, name, and address that is provided in the card. It will also update e-mail and URLs), provides information about book, movies, music (scan up the book cover and Google Lens will automatically provide you with the books review via the web, it will also tell about the movie, music albums, games just by capturing its cover), details of art and architecture (capture a fascinating building/architecture and get to know the detail of its construction, architect, also get to know about the art pieces at the museum and many more), scan products (scan the bar codes, QR codes), recognize images (get to know the beautiful flower at your neighbor's garden, just by capturing it and let Google Lens identify which flower it is, it also tells the animal that is being clicked). The first thing to do is open Google Photos app on your smartphone. In it, open up any photo and, if your phone does have an update, you can see the Google Lens icon at the bottom of the image—it will be square icon with a circle inside it. Next, click on the Lens icon while opening a photo. Once you have opened, you will see that dots are displayed on the phone's screen as the phone checks the objects in the photo. Google Assistant will immediately pop up and provide information about the objects contained in the photo. One can also use Google Assistant to quest for more knowledge through Google Search if one wants to discover more. While testing it, the AI enriched service correctly recognized and identified the Prague Castle, St. Vitus Cathedral, Flatiron Building in New York, but hesitated to identify Shanghai's Oriental Pearl Radio and a simple TV Tower (identified it as a "skyscraper") through the Smartphone Pixel 2.

Another application called the **CamFind** promises to do similar visual search like Google. It is powered with Cloud-Sight Image Recognition API. Just simply take a picture of anything and CamFind can help you provide the exact, similar results within a few seconds. A search result contains relevant similar images, videos, shopping results from locals, and also a huge section of web results. It also provides an option to save the results in your profile so that you can share it with your friends and family.

10.3.2.5 GOOGLE GLUCOSE-SENSING CONTACT LENSES

In the present lifestyle, diabetes is a common outbreak that seems to be growing in the world, and also the number of people who need to take the daily tests of pricking the finger to monitor the blood sugar level.[21] Google has announced that it will be working on the implementation of the "smart" contact lens that can measure the quantity of glucose content in tears. If this works successfully, this lens would be the commencement of a finger prick-free prospect.

The collaborators Brian Otis and Babak Parviz explained the technology as a normal soft contact lens material that contains a tiny glucose sensor and a small wireless chip: The testing protocols will tend to generate a reading once per second. Plans are also to include micro-LED lights that would turn up when there is high level of glucose (if the level crosses the limit) to indicate alarming situations. So far, Google is looking for partners that would take up the project to make the lens. Other companies are looking for an alternative of finger pricking which includes measuring glucose through breath and saliva. A company known as Freedom Meditech is developing a mini device which can help to measure glucose levels with the help of eye. In the United States, almost 8% are affected with diabetes. There is always space for innovation, a smart contact lens or even saliva monitor will make it considerably easy to check out for the raise in blood sugar levels. As of now the company has halted to develop the contact lens due to insufficient consistency in the tasks, and exact correlation between tear and blood glucose levels. Let us wait until they find out that.

10.3.2.6 GOOGLE'S DEEP DREAM

One of the leading applications of modern machine learning is associated with images;[22] the objective is to train computers in analyzing, classifying, and also altering the different types of images. In the year 2016, Google's Deep Dream software (human–AI collaboration) became famous by creating a chain of terrifying, nightmare-inducing images. Image and video classification is not so innovative anymore, number of blogs, forums have made the work easier by making image classification reasonably available. Images are fed by the researchers and artists to produce an art. Furthermore, the computer starts to work by creating pictures of spiral vortexes and psychedelic alike towers, morphing animal faces and other beautiful landscapes. Now everyone's turning into an artist.

10.3.3 GAMING

The idea of spending long hours playing video games is usually not advisable for humans to augment their intellect, the sensible 3D graphics and environment of videogames have made videogames the ideal learning tool. Some of the AI-inspired videogames are Assassin's Creed, Microsoft's Project Malmo, Grand Theft Auto, PUBG, and many more. AI has been programmed in a different way for gaming. Starting from the use of finite state machines in the Pac-Man game, and to the A* algorithm in the Mario game, the development has been immense using the deep learning algorithms, computer systems will be able to manage cognitive process in games. An incredible machine learning trick clears up pixelated images and videos, Magic Pony algorithms sharpen up a pixelated character or object that is to generate high-quality computer-game graphics. A lot of development in technology such as consoles, cloud/connectedness, ultrapowerful graphics cards, virtual reality, or headsets are aiding AI to provide an even more attractive and impressive environment in which virtual characters show human behaviors and intelligence.

In the early 2017, Children's Childhood Anxiety Reduction through Innovation and Technology launched an interactive video game known as "Sevo the Dragon," which projects on the Bedside Entertainment and Relaxation Theater screen and turns it into a relaxing game in the administration of anesthesia, so the little patients have interesting fun to do while taking in medicine through a mask.[23] Caruso and Rodriguez are at present collaborating with Silicon Valley-based software engineers to form original virtual reality (VR) content that is particularly customized to pediatric patients. Spaceburgers, the first game, was created in joint working with Juno Virtual Reality, allowing the kids to listen to relaxing music as they are flying to outer space and zap the incoming objects.

Researchers are at the moment investigating how virtual reality has an impact on the pain and anxiety levels during medical procedures and as of now, the results are promising. The little ones who are engaged with VR are likely to be more cooperative, less scared, and experience less soreness/pain during procedures.

VR distraction treatment is being used for kids at Packard Children's of age 6 in treating particular areas like the emergency department, and it will also make an entry into the labor ward and delivery unit. Children in particular should not be scared of visiting a Doctor to get injected with antibiotics. Certain bad experiences cause phobias that last till the end, now, patients will get injected by simply wearing VR glass and feel painless.

10.3.4 SONY'S CONTACT LENS AND STORAGE MEDIUM

Do you know how a Human eye captures a scene or image, the resolution of human eye is 576 MP.[24] Although for knowledge human eye was taken for example, there also exists an application where a lens has been fitted on the human iris which on the blink of the eye clicks scenes/images and also stores in the storage medium. Developed by Sony, named as "Contact Lens and Storage Medium," despite its apparently unbelievably small dimension, a completely developed imaging device, including a lens, data storage, an imaging sensor, and a wireless communication component that captures images at the blink of the eye and also stores it.

According to the method, the lenses will first record images when they are worn on as a lens on an eyeball. The data will be recorded on their own storage units provided. And how does it records or captures photos and videos? It just happens with a "wink."

Sony's patent clarifies that "in a case where prearranged eyelid closing of an eyelid that is in contact with the lens unit is detected, the recording control unit records the captured image which was snapped by the image pickup unit that is in the storage medium." That means the lenses knows the human blinking patterns and can recognize if you are blinking out of need or blinking a little longer to indicate that you would like to start the video recording. Sony says one would only require wearing of a smart contact lens in one eye. The application says the contact lenses could be customized as hard or soft, depending on the user's preference. As the common saying goes, Power is in the hands but now we need to change it as "Power is in our Eyes" which is offered wirelessly. Imagine capturing something fascinating on the go or even witnessing crime occurring on the way. We should also see the development of our hand-held devices, where face lock has been incorporated on the devices.

Though it would be great for parents to record their kids' on-stage performance without having a need to hold up a camera or for a new bride to record her wedding vows from her own eye, other usages of the contact lenses may bring some worry in privacy concerns.

Gizbot's Rohit Arora suggests that "While this really sounds futuristic, at the same time we believe that Sony's contact lenses can cause a breach of an individual's privacy. The person would never know that someone with such sophisticated tech can record his/her activities with just a blink of an eye."

Another tech writer William McKinney brings up the exasperation that can come with the new communications medium. "If we start viewing the internet in our glasses and contacts, next-generation advertising would come straight to our faces," writes for the Edgy Labs.

Although for those who are eager to check out the contact lenses will most likely have to wait for at least a year or two, as no release date has been announced yet.

10.3.5 OCUMETICS BIONIC LENS

In May 2015, a Canadian ophthalmologist announced that he has created a bionic lens that could correct a person's vision for life, effectively resulting in vision three times better than 20/20.[25] The product came into life after eight years of research and about $3 million in funding, the Ocumetics Bionic Lens is said to require a painless 8-min in-office procedure that requires no anesthesia.

The researcher folds up the custom-made lens like a tiny taco (a Mexican dish) to fit it into a saline-filled syringe. Then, uses the syringe to place it in the eye through a supersmall incision, then leaves it there to unravel over about 10 s. And it is finally done.

The bionic lens is made to replace our natural eyes; the process may get rid of any risk of cataracts in the future. As cataracts may release chemicals that raise the risk of glaucoma and other issues, it also helps in protecting the overall eye health.

10.3.6 SECURITY

Security camera checking is regularly conducted through human working. Humans may have trouble in tracking multiple monitors at the same time.

Like other applications of AI here, AI can be trained using supervised exercises, security algorithms, and so on to take input from security cameras. Eventually, they can identify potential threats and warn human security officers to investigate further.

Currently, they are pretty limited in what they can perceive as a threat. *Wired* detailed that they can see flashes of color, for instance, that may indicate an intruder or someone loitering where they should not be doing it. More sophisticated misbehaviors like identifying potential shoplifters are far beyond its capability at present. This is likely to change very quickly as the technology improves.

As of now, AI identifies criminals with an accuracy rate of around 89.5%. It achieves this with the help of machine and vision algorithms. AI uses photos of suspects and real criminals without facial hair. AI not only identifies the criminals with a high success rate but also managed to provide typical facial features that might indicate the person is less than law abiding.

Such determinations definitely brought up moral concerns, particularly whether or not it was just to associate these features with criminal activity. Some of the traits found by the AI included:

- inner corner distances of their eyes
- specific lip curvatures, and
- nose–mouth angles.

10.3.7 FINDING A MISSING PERSON

Applications of AI could be used for disaster events in finding isolated survivors. Formally, one would either have to go out in person and search on foot or look at aerial footage of the disaster areas.

Checking each and every CCTV footage and photo is very time consuming and any lost time could have been utilized essentially to help someone survive before death.

Drones are by now in use to offer on-time footage of catastrophe areas which still rely on humans to examine the captured footage. AI authorizes the assessment of large amounts of data, photos, and CCTV clips/footage to help in finding the missing person, sometimes in a short time of less than 2 h. It can even find piles of debris in flooded areas that may have rapt victims in them. AI is also used to analyze social networking sites like Twitter to know about who went missing during the disaster.

10.3.8 WILDLIFE MAINTENANCE

Wildlife preservation is particularly difficult, especially when attempting to analyze population sizes or in tracking animals.[26] Scientists and researchers cannot probably track every animal or tag them all with GPS devices.

- A team from Chicago has effectively implemented a form of AI, developed by Wildbrook.org, to perform this task for them. AI inspected photos are uploaded online and with a smart use of algorithms, it analyses each photograph and looks out for unique markings.
- It also helps in tracking the habitat ranges from GPS coordinates provided through each photo, guess the animal's age and even determine its gender.
- The team conducted an enormous campaign in 2015 that managed to find out that lions were killing too many baby Grevy's Zebra in Kenya. This alerted the local officials to change the existing lion administration program.

10.3.9 OTHER FINDINGS

We all must have come across "The dress color illusion" that made rounds over the internet; the controversy still persists whether the dress in the image is blue–black or white–gold. Many people come to a conclusion that for the left brain users the image seems to be as blue–black and those who use right brain, the image seems to be gold–white. But what actually is happening in a single image? How do these variations occur? The answer for this is a color illusion. Color illusions are the images in which the object's contiguous color tricks the eye of the viewer into interpreting an incorrect color. What actually makes the dress appear in such a way is that our eye either slashes out the blue so we see the white and gold, or it slashes out the gold so we see the blue and black. But why would the eye behave in this way? Another common explanation to it is human beings progress to see in daylight, but that daylight changes the color of the whole thing we see with our eyes. Human eyes try to atone for that chromatic bias of color of daylight. We generally get to see objects or scenes because of the light that gets reflected. When we observe something, light enters the eye with differing wavelengths which match up to different colors. The light initially hits the retina at the back of eye, the region where the pigments shoot signals to brain's region which helps to process the signals into an image. The brain in turn finds out the color light that bounces off the object to eyes that is being visualized by subtracting the color from the original color of the object. Bevil Conway, a neuroscientist who does research in color and vision at the Wellesley College in Massachusetts, declared "Most people will see the blue on white background as blue. But on the black background, some may see it as white."

A Chinese surveillance company named Watrix has developed a new system for **gait recognition** which can simply identify people as far as 50 m away, as reported by the company's CEO Huang Yongzhen.[27] It will analyze how individuals carry out themselves and, as of now it is not capable of real-time recognition still, the company claims it will definitely analyze and search in an hour's time of clipping in 10 min with an assured accuracy rate of 94 pc. Currently, there are about 170 million CCTV cameras in the country and the numbers are going to grow, with more than 400 million cameras to be installed by 2020.

Our homes are flattering smarter day by day. Many smart and elegant devices are now made to learn our behavior patterns and can also help us save money a lot of money. Thermostats and building supervision systems can assist to computerize building heating and cooling. In consequence, they get trained and then predict when to switch boiler on or off for finest comfort, while keeping update with the outside weather circumstances. Future ovens, pointless to say, real chefs, will keep your food ready when you get back home from work. Lighting is yet another model of a home appliance which gets the AI alteration. They would be used to set evasion and fondness for lights around homes. They will also be able to dim lights while watching TV. The outlook is certainly bright with AI, or dim, or just plain.

There is another variety of applications of AI. Much simpler while comparing to other AI things like Spotify, Pandora, and Netflix, as they provide a helpful provision to the users. It will recommend music, movies, based on the viewers' interests or watching history. Rather more or less alike predictive purchasing, one's "liking" in stuff could make you spend a lot. The AI systems would observe the choices and insert them in a learning algorithm. From which, it recommends to similar things one would have liked. It mostly predicts based on the input given or recorded. As invasive as it might sound, it can also help you determine something that may become your new preferred thing.

With the growing Android phone usage, applications are created every day, even for a smallest to-do list there is an app. Children used to enjoy their favorite video games, just by holding a joystick and rolling it over to make a move. But with AI's gesture recording, one can actually make a movement and that reflects in the characters movement virtually. Gesture movement has become a varying trend, where one can click a photo with a hand gesture (wave), get a picture captured with a smile. A person's approximate age and gender are detected while looking onto the front camera, how does it happen? The **Xiaomi's Camera Software** and **Microsoft's age guessing website** follows the face detection technique where attributes such as emotion, wrinkles, pose, facial hair, smile, and other 27 features for every face input are identified. In the **android phone camera,** one can also change or add effects to the image, make it grayscale, wrinkle-free, make it sun-kissed, add snow effects, and lot more. All of these are possible only with the help of AI's quickly developing usage into real-world products and its advancements for a better lifestyle.

With the extreme advancements, the normal lifestyle is changing, e-commerce applications and websites (Zomato, Swiggy, Amazon, Swiggy, Myntra, LensKart, Ola, Uber) have made people come closer to the digital world, where people on the click order food; buy clothing based on the fittings on a model, select eye wear just by uploading a photo and lots more, book an auto or a cab based on our preference, also check the arrival timings and travel hassle-free. Exploring how each and every product and the usage of AI into it has to be known so that one becomes aware of it without reading the manual.

Techniques helping in learning, automated reasoning, and perception have made a common entry in everyday lives. You can ask whatever you need to from Apple's voice assistant **Siri,** Microsoft's **Cortana,** Google's **Google Assistant,** Amazon's **Alexa,** and many other to help us. With advancing AI systems, there may be a time were people would just say "Transfer amount to my House Owner" and automatically amount would be transferred to the house owner's account number.

10.4 DISCUSSIONS AND LIMITATIONS

The approach of how image recognition works, normally, involves the making of a neural network that processes every single pixels of an image. Researchers feed in with these networks as countless prelabeled images as possible, so as to "teach/train" them on how to recognize similar images when given at a later stage.

Nowadays, we are enclosed with using technology, from the smart gadgets present in our houses, smartphones in purses and pockets, computers on our tables, and the routers that connect us to the world (internet). In each one of the technologies used, the functional architectures proceeded accurately, all thanks to the base engineering ideology they were built upon mathematics, physics, electrical science, computer, information technology and software engineering, and so on and above all these domains—many years, and decades, of statistical testing and quality guarantee.

It is significant to consider that deep learning models require a huge amount of data to train a first model (to attain high accurate results and not to produce overfitting, keeping in mind that sequential tasks can learn from the transmit learning), and that eventually without a deep understanding of what is actually happening inside a "deep neural network," it is not practically nor hypothetically sensible to build technological solutions that would be feasible on the long run.

At this point, AI is still like a growing child. Computer vision is giving it the development of sight, but that does not come with an existing understanding of the world. For which, AI has to undergo training just like kids do. If you teach a kid by drawing a number or showing a letter plenty times, it will learn and understand ways to recognize the number or letter. Amazingly, many kids can instantly recognize letters and numbers even upside down once they have learned it correctly. Our biological neural networks are logically good at understanding visual information even if the image that is being processed is not exactly how it is expected to be. Let AI grow to a larger extent that would make individuals wonder what work they need to do as AI would completely take up all of the human's responsibility.

10.5 CONCLUSION

AI has influenced our lives in such a way that even without our liking we need to incorporate it in our daily lives. Recurring tasks that are tedious in performing can be carried out with the intelligent machines. Machines processing speed is quite faster than the thinking capacity of humans, and hence machines can be employed in doing multitasking and also be engaged to do dangerous operations. Their working, modules, and other parameters can be adjusted or changed based on the requirements.

AI with imaging techniques is helpful in health care to provide a faster diagnostic analysis response to identify, recognize, and also classify the affected disease. AI in business helps to predict shares through graphs from already existing records. AI also provides services such as finding out the missing person, recognizing facial features, adjusting photos quality to know about the scripts in ancient photographs, and many more. Laborious tasks and hard work required to achieve the prescribed task are done easily with AI techniques and Systems. Since these AI systems and techniques do not wear out easily (breakdown or disfigured), the sustainability time is high. Techniques helping in learning, automated reasoning, and perception have made a common entry in everyday lives. Let AI learn, train itself, and then help us to achieve great heights.

KEYWORDS

- **artificial intelligence**
- **digital world**
- **neural network**
- **computer vision**

REFERENCES

1. Ray, S. History of AI; 2018, August 11. https://towardsdatascience.com/history-of-ai-484a86fc16ef.
2. Olanfenwa, M. Train Image Recognition AI with 5 Lines of Code; 2018, July 20. Retrieved from https://towardsdatascience.com/train-image-recognition-ai-with-5-lines-of-code-8ed0bdd8d9ba.
3. Fogg, A. A History of Deep Learning; 2018, May 30. Retrieved from https://www.import.io/post/history-of-deep-learning/.
4. Melissa, T. Reading the Readers: Modelling Complex Humanities Processes to Build Cognitive Systems. *Literary Linguist. Comput.* **2005,** *20* (1), 41–59.
5. Computer Technology Helps Radiologists Spot Overlooked Small Breast Cancers; 2000, Oct. 1. Retrieved from https://www.cancernetwork.com/articles/computer-technology-helps-radiologists-spot-overlooked-small-breast-cancers.
6. Fei-Fei, L. ImageNet: Crowdsourcing, Benchmarking & Other Cool Things. *CMU VASC Seminar* **2010,** *16*, 18–25.
7. Markoff, J. How Many Computers to Identify a Cat? 16,000; 2012, June 25. Retrieved from https://www.nytimes.com/2012/06/26/technology/in-a-big-network-of-computers-evidence-of-machine-learning.html.
8. Taigman, Y.; Yang, M.; Ranzato, M. A.; Wolf, L. Deepface: Closing the Gap to Human-Level Performance in Face Verification. In *Proceedings of the IEEE Conference on Computer Vision and Pattern Recognition*, 2014; pp 1701–1708.

9. Condliffe, J. AI Has Beaten Humans at Lip-reading; 2016, November 21. Retrieved from https://www.technologyreview.com/s/602949/ai-has-beaten-humans-at-lip-reading/.
10. Brindale, B. How Does Google Maps Predict Traffic? Retrieved from https://electronics.howstuffworks.com/how-does-google-maps-predict-traffic.htm.
11. Glover, J. An Introduction to Generative Adversarial Networks (with Code in TensorFlow); 2016, August 24. Retrieved from http://blog.aylien.com/introduction-generative-adversarial-networks-code-tensorflow/.
12. Marr, B. *Data Strategy: How to Profit from a World of Big Data, Analytics and the Internet of Things*; Kogan Page Publishers, 2017.
13. Retrieved from https://www.butterflynetwork.com.
14. Retrieved from https://clarifai.com.
15. Retrieved from https://support.apple.com/en-in/HT208955.
16. Retrieved from https://echonous.com/en_us/
17. Retrieved from https://www.infervision.com/en
18. Retrieved from https://www.makeuseof.com/tag/technology-explained-google-maps-work/.
19. Retrieved from https://electronics.howstuffworks.com/gadgets/other-gadgets/project-glass.htm.
20. Retrieved from https://lens.google.com
21. Retrieved from https://www.fastcompany.com/3025081/google-is-working-on-glucose-sensing-contact-lenses-for-diabetics.
22. Retrieved from https://deepdreamgenerator.com.
23. Retrieved from https://www.stanfordchildrens.org/en/innovation/chariot
24. Retrieved from https://curiosity.com/topics/smart-contact-lenses-could-record-every-thing-you-see-curiosity/.
25. Retrieved from http://ocumetics.com/
26. Retrieved from https://interestingengineering.com/17-everyday-applications-of-artificial-intelligence-in-2017.
27. Retrieved from https://www.siliconrepublic.com/enterprise/china-gait-recognition-cctv.

CHAPTER 11

A Novel N-Average Wavelet Algorithm for a Voice-Based Wheel Chair

E. CHANDRA*

Department of Computer Science, Bharathiar University, Coimbatore 641046, India

**E-mail: crcspeech@gmail.com*

ABSTRACT

For the past few decades, speech recognition has been an interesting field of research. Speech recognition has been the best means of communication and it is probably the easy means of communication too. When compared to the other input devices like keyboard and mouse, it is very easy to input and less time-consuming. The speech recognition is achieved through many means of algorithms and architectures. In the field of signal processing and machine learning, the neural network plays an important role. In speech recognition of machines, deep learning algorithms are applied. To train the machines with the consistent speech corpora, classifiers of neural networks like recurrent neural network (RNN), deep recurrent neural network (DRNN), and deep belief network (DBN) are applied. The ultimate target of the deep learning process is to design a machine which could react like a human being, that is, it should learn things like sense organs of humans, remember things like a human brain, learn the action, and recognize things at the times of need. Above all, a language model is very essential for a speech recognition system. The generation of regular grammar, words, and language syntax are formed by a language model and is very helpful to identify the recognition of words easier. The novelty of the work focuses on the generation of a language model for Tamil speech recognition system for making use of a wheelchair for people suffering from permanent or temporary or birth disability in order to drive a wheelchair using Tamil voice commands for physically handicapped and aged people. This chapter proposes two approaches for speech recognition. Zero cross rate (ZCR) is used for speech feature extraction in the first approach and secondly, DBN is applied for automatic speech recognition (ASR). For different types of speaker's identification, DBN is adapted as the effective acoustic model. A novel N-average wavelet algorithm is applied to extract the speech features in the second approach. Adaptive neuro-fuzzy inference system is used to train the noiseless data. The adaptive network-based fuzzy inference system is a type of artificial neural network that is developed based on the inference system which was developed in the 1990s called Takagi–Sugeno fuzzy inference system. By implementing this proposed system more accurate result has been achieved with an accuracy of 99% for Tamil commands and with a maximum elapsed time of about 0.5 s.

11.1 INTRODUCTION

Machines try to understand the words and sentences spoken by humans. The process of identifying words and phrases by the ability of a machine or program in spoken language and convert them into an understandable form is called automatic speech recognition (ASR). Speech is recognized by considering its various attributes such as its energy, trajectory of utterance, speaking mode, and different speaker utterance. The same phoneme is made completely complex by different speakers, speaking mode, and context—the fundamental unit of speech segments. Segmenting is the process of subdividing the word into smaller units of speech called segmentation. Speech recognition of limited simple words can be done by the traditional method, whereas when we have to recognize more words an efficient method is required. Deep belief network (DBN) can be used for such purpose. In ASR, converting spoken words into a text is still a hazards task due to the high dynamic variation in speech signals. Neural network is rarely successful for continuous recognition tasks so deep learning comes as a rescue for continuous recognition of words. The first phase of speech recognition is preprocessing which records the speech signal as analog signals, with respect to time. The digital processing involves sampling of continuous analog signals into digital valued signals.

In speech recognition systems, the preprocessing is used to increase the overall efficiency of the recognition performance by extracting the features of the speech signals and classification of different stages of the speech performance. In the preprocessing step, the voices are sampled, filtered for noises, and windowed for the next process. At the end of the preprocessing, the voices are sampled and made available for the recognition process. Before the feature extraction stage, the voices are compressed, normalized, and filtered. After the feature extraction, the speech signals are characterized according to the feature extracted. The extracted features are sent to the DBN algorithms and compared with the traditional method and finally conclude that the deep belief algorithm has a better performance. Hidden Markov models are combined with neural networks for an efficient speech recognition system.[1,2] Deep Feed forward networks are used for improving the acoustic model which has gained great attention in the past years. The pattern analysis and unsupervised learning are done by the exploitation of machine learning techniques with different hierarchical architectures. Deep learning is an important method for learning machine techniques, which contains many stages of information processing in the priority of architectures. The deep learning algorithm has become successful in dealing with the artificial intelligence (AI) problems and speech recognition processing. Deep learning is used to compute hierarchical data and observational data from a smaller level to a bigger one. The training of data in a large amount and the capability to represent the acoustic features has inspired the researchers to use the deep learning techniques in speech recognition. Deep learning algorithms have proved to solve complex problems with greater efficiency. AI is a field in which expects a model which solves the complex task with high efficiency and the approaches have turned to be a successful one. There are many different types of deep learning architectures available with good learning capacity and with excellent performance. Among all the deep learning architectures the Convolution Neural Network and DBN seem to be the best.

Speech recognition is the conversion process of spoken words into basic language units or phonemes. The main purpose of using a neural network is to recognize the pattern-matching abilities on the speech signal. Due to the increased computational capacity in recent years, deep learning algorithms play an important role in speech recognition. Deep learning allows estimating the feature segments in a natural and effective way.[1] The voice signals are given as input for preprocessing. The input data are filtered for noise and silence. These filtered data are divided into segmentation using blind segmentation. Segmentation is the process of uniquely identifying meaningful speech units such as voice, phonemes, boundary values, syllables, words, and phrases, etc., and they are processed in order to extract the features of the voices.[2] In the speech recognition process, segmentation helps us to reduce the complexity of computation and the memory size of the large vocabulary words. For evaluation of segmentation systems, the most important measures used, are % false detection (FD) and % false rate (FR) which are calculated as shown in eqs 11.1 and 11.2.[2]

$$\%\,\text{False Detection}\,(\text{FD}) = \frac{\text{False Detection}}{\text{Total Amount of Detection}} \quad (11.1)$$

$$\%\,\text{False Rate}\,(\text{FR}) = \frac{\text{Missed Detection}}{\text{Total Amount of True Change}}. \quad (11.2)$$

Accuracy is determined by the segmentation method, F measure is defined as shown in:

$$\text{F} = \frac{2 \times (1 - \text{FD}) \times (1 - \text{FR})}{2 - \text{FD} - \text{FR}}.$$

11.2 PROPOSED SYSTEM

11.2.1 FEATURE EXTRACTION USING ZERO CROSSING RATE (ZCR)

Next process is to extract the speech extraction using ZCR. Several processes of modification takes place to receive the different characteristics of speech at different levels such as articulation, intonation, speech pitch, semantic, speech rate, pronunciation, linguistic, and various other changes. The variations that happen in the above transformations are considered to appear as the differences in the acoustic properties of the speech signals.[2] A well-formed recognition system is to be determined for representing the information of the speech signal. This transformation of the signal will help us to identify the signal in different domains. The ZCR is one of the feature extraction methods in speech processing to calculate how many times the speech waveform has crossed the zero axis. Below is the general formula for ZCR. The feature is extracted by the following steps:

- Frame the speech signal into segments.
- Apply the ZCR to the segmented signals.
- Filter the feature that reaches the signal spectrum.

- Calculate the feature spectrum by applying the below formula.
- $ZCR = \frac{1}{2N}\sum_{n=1}^{N}\left|\text{sign}(x[n]) - \text{sign}(x[n-1])\right|$.
- If the ZCR value is less than the feature is taken as voiced data.

11.2.2 DEEP BELIEF NETWORK

Deep learning a branch of ML is really doing justice to all those valuable data floating around in this universe and processing it efficiently to help us reach to some rational conclusions in the field of speech recognition, image recognition, NLP, healthcare, financial sector, etc. Normally, a deep learning architecture consists of the following three layers. The inputs are given through the input node, the intermediate nodes called the hidden node, and to display the output through the output node.

Deep learning is a structured network which consists of deep networks of different topologies. Neural networks are a field that has been in existence for quite a lot of time. It has enormous layers of networks for the development of particular functions according to the required constraints and layers of networks are built for providing the feature extraction that has made the neural network a more practical to use.[4] More and more flexible layers are added to provide the weights and interconnections between the layers for their better performance. For the different spectrum of problems, deep learning approach with different topologies is applied. The architectures and algorithms that are applied and used in deep learning are wide and varied in numbers.

In the above section, we explore various types of deep learning architectures which are in use for the past few decades. The recurrent neural network (RNN) is one of the basic network architectures from which all the other deep learning architectures are built. A RNN makes a back propagation to learn about the previous layers. This is the main difference between the normal multilayer networks with that of the recurrent network. A recurrent network has connections that can feedback into previous layers.[5]

The basic attributes of DBNs are:

- The input nodes are given with a set of featured inputs.
- The hidden nodes are normally self-adjusting according to the complexity.
- The output nodes are trained with the given data.

In general, the performance of a speech recognition system is measured in terms of recognition rate and word error rate.

$$\text{Recognition rate}\ (\text{RR}) = \frac{\text{Number of words recognized} \times 100}{\text{A total number of words spoken}}$$

$$\text{Word error rate (WR)} = 100 - \text{Recognition rate.}$$

Furthermore, a DBN has been implemented for ASR. The speech recognition performance is evaluated on three different types of speech recognition systems, namely, Gaussian Mixer Model–Hidden Markov Model, Deep Neural Network-Hidden Markov Model, and DBN with different corpus data set in terms of word error rate and the evaluation has been made. The results have shown that the DBN-based speech recognition system excels over the other two speech recognition systems.

11.2.3 SPEECH RECOGNITION USING ANFIS AND N-AVERAGE WAVELET METHOD

In the N-average wavelet algorithm approach, several families of wavelets that are proven to be especially useful are included in this method. There are 53 types of wavelets available such as Haar, Mexican hat, Daubechies, Coiflets, Biorthogonal, Morlet, Meyer, Symlets, Other Real Wavelets, and Complex Wavelets. Apart from the 53 wavelets, we have selected Haar, db1, sym3, coif1, bior3.1, rbio3.1 wavelets for the better feature values. From the 10 voice signal, energy feature is extracted for all 6 wavelets. The feature values are normalized then ranked. The normalized feature values are sending to the adaptive neuro-fuzzy interference system (ANFIS) algorithms and then the performance is analyzed.[6]

11.2.4 FEATURE EXTRACTION USING DIFFERENT WAVELETS

Feature extraction is the most relevant type of classification that extracts features from the raw data and derives the information which is most relevant for classification purpose, in terms of minimizing the pattern variability within the classes while enhancing the pattern variability between the classes.[7] The dimensionality of data is reduced during the feature extraction process. A good feature extraction scheme enhances the feature of the input data which separates the different pattern classes. This becomes very essential due to the limitation in the technical features like memory and computation time.[8]

11.2.5 FEATURE SELECTION USING N-AVERAGE WAVELET ALGORITHM

Step 1: Acquiring energy feature values from each audio signal.[9]

Step 2: Normalizing the feature values by dividing the individual values by the maximum value of each feature.

Let wavelet features: F1, F2, F3....F53. n = Number of audio signals considered for training.

$$\frac{F1(1,2...n)}{MaxF1(1,2...n)}$$

Step 3: Taking minimum and maximum of individual feature values. F_{min}=Min ($F1$ (1,2…n)) F_{ma}=Max ($F1$ (1,2…n))

Probably the max ($F1$ (1,2…n)) will be 1 for all the feature values.

Step 4: Taking the difference between the F_{min} and $F_{max.}$ $F_{Diff=}$ $F_{min} \sim F_{max}$

Step 5: Arranging the F_{Diff} for all the 53 features in largest to smallest order and taking the first five maximum features.

11.2.6 THE FUNCTION OF THE ADAPTIVE NETWORK FUZZY INFERENCE SYSTEM (ANFIS)

The neural network and fuzzy logic are the sophisticated interrelated tools in building artificial intelligent systems. The low-level structures are performed well with the raw data in the neural network architecture. With the linguistics knowledge acquired from the experts, the fuzzy logic deals with the higher level of reasoning ability. The neural networks have the learning ability to attain the knowledge just like humans and fuzzy systems have the ability to explain the knowledge it has attained. This makes both the concepts to be combined together and form integrated concepts called neuro-fuzzy systems that can work parallelly with both the concepts. Our adaptive neuro-fuzzy inference system is a hybrid learning algorithm that combines the gradient descent and least square estimators. A Sugeno model is an advanced model which is applied in our proposed design.

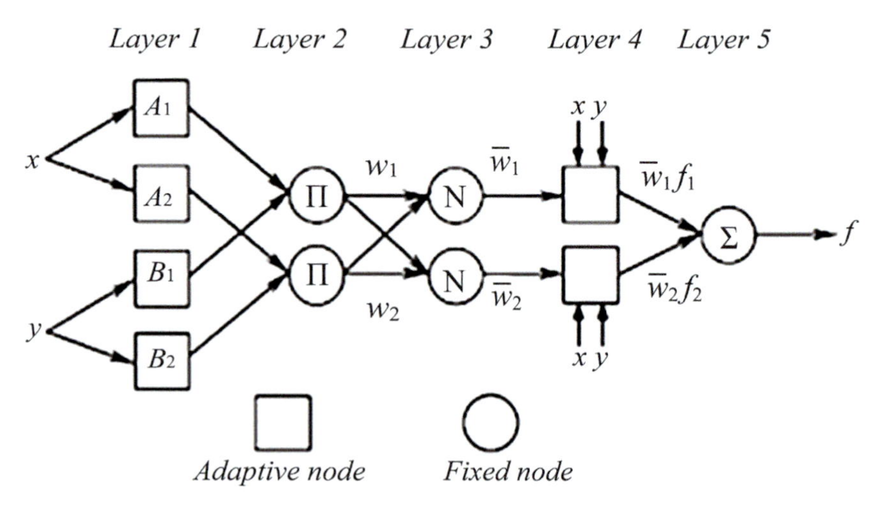

FIGURE 11.1 Takagi–Sugeno ANFIS model.

The steps followed in the ANFIS Sugeno model are:

- Send the selected features to the layers of ANFIS.
- It consists of six layers.
- The input is passed through the different layers and trained.
- The new inputs are compared with trained data for efficiency.
- The efficiency found to be 99% for Tamil commands and with a maximum elapsed time of about 0.5 s.
- For learning algorithm in an adaptive network, ANFIS architecture uses supervised learning that is similar to the Takagi–Sugeno model of fuzzy inference system. The function is similar to the model of the Takagi–Sugeno model. For simplicity, assume that there are two inputs x and y, and one output f. Two rules were used in the method of "If-Then" for Takagi–Sugeno model as follows:

1. Rule 1: If a is x1 and b is y1 Then f1 $= p_{1\,a+}\, q_{1\,a+}\, r_1$
2. Rule 2 : If a is x2 and b is y2 Then f2 $= p_2 a + q_2 a + r_2$

11.3 CONCLUSION

A new control system of a voice controlled wheelchair using AI algorithm is proposed. The machines are given instruction through the voice signals and it has become inevitable for communication in this current state of technology. It is robust to find the perfect match from the continuous voice recognition algorithm for the given pattern of voice. In the proposed work, wavelets are used for the feature extraction of the input voice signal; here energy feature is extracted from the given input signal. It has proved a more effective feature among all other feature sets. To achieve a high and competitive performance of the voice recognition, adaptive neuro-fuzzy inference system is proposed. The proposed work brings a novel approach to train the Tamil voice signal and move the wheelchair based on the commands by using ANFIS.

Better or comparable results are obtained with this proposed model. Though the number of data sets used for system modeling in our proposed system is less as when compared to that used in traditional or conventional methods. The real-world systems build an ANFIS model which will be very useful for solving robust tasks like pattern recognition and speech recognition. ANFIS models are robustness against the failure of single units. It can design its own network architecture by learning from data.

KEYWORDS

- **wavelet algorithm**
- **voice-based wheel chair**
- **automatic speech recognition**
- **artificial intelligence**
- **speech recognition**

REFERENCES

1. Halager, A. Speech Recognition using Deep Learning. (IJCSIT) *Int. J. Comput. Sci. Information Technol.* **2015,** *6*(3), 3206–3209. www.ijcsit.com 3206.
2. Räsänen, O. *Speech Segmentation and Clustering Methods for a New Speech Recognition Architecture*; Espoo, 2007.
3. Jayasankar, T.; Thangarajan, R.; Arputha Vijaya Selvi, J. Automatic Continuous Speech Segmentation to Improve Tamil Text-to-Speech Synthesis. *Int. J. Comput. Appl.* **2011,** *25*(1), 31–36.
4. Ganesh, A. A.; Ravichandran, E. C. Syllable Based Continuous Speech Recognizer with Varied Length Maximum Likelihood Character Segmentation, International

Conference on Advances in Computing, Communications and Informatics, 2013; pp 935–940.

5. Hinton, G. E.; Osindero, S.; Teh, Y. W. A Fast Learning Algorithm for Deep Belief Nets. *Neural Comput.* **2006,** *18*, 1527–1554.
6. McAfee, L. Document Classification using Deep Belief Nets, CS 224n, 6/4/08.
7. Xiaohu, W.; Feng, Qi; Haijiao, Y.; Jun, W.; Jianhua; S.; Zongqiang, C.; Haiyang, Xi. Wavelet and Adaptive Neuro-Fuzzy Inference System Conjunction Model for Groundwater Level Predicting in a Coastal Aquifer. *Neural Comput. Appl.* **2015,** *26.* 10.1007/s00521-014-1794-7.
8. Jang, J. S. R. Fuzzy Modeling Using Generalized Neural Networks and Kalman Filter Algorithm. In Proceedings of the AAAI'91 Proceedings of the Ninth National Conference on Artificial Intelligence, Anaheim, CA, USA, 14–19 July 1991.
9. Amid, S.; Gundoshmian, T. M. Prediction of Output Energies for Broiler Production using Linear Regression, ANN (MLP, RBF), and ANFIS Models. *Environ. Prog. Sustain. Energy* **2017,** *36*, 577–585.
10. Banumathi, et.al. Speech Recognition of Continuous Tamil phoneme using DBN. *Int. J. Innov. Res. Comput. Commun. Eng.* **2016,** *4*(7).

CHAPTER 12

Automated Diagnosis of Heart Disease Using Artificial Intelligence

E. UDAYAKUMAR[1], S. BALAMURUGAN[2*], and P. VETRIVELAN[3]

[1]*Department of ECE, KIT-Kalaignarkarunanidhi Institute of Technology, Coimbatore, India*

[2]*QUANTS IS & CS, Coimbatore, India*

[3]*Department of ECE, PSG Institute of Technology and Applied Research, Coimbatore, India*

[*]*Corresponding author. E-mail: sbnbala@gmail.com*

ABSTRACT

Cardiovascular malady is one of the major reasons for death around the world. The level of early demise due to coronary illness happens at a pace of 4% in high-pay nations and 42% in low-salary nations. This shows the significance of anticipating coronary illness at the beginning period. In the restorative field, the finding of coronary illness is the most troublesome undertaking to be finished. It relies upon the cautious examination of various clinical information of the patient by restorative specialists and it is a muddled procedure. An expectation framework for coronary illness is proposed utilizing adaptive neuro-fuzzy interference system (ANFIS). This framework acknowledges different clinical highlights as contribution as name, age, sex, blood pressure (systolic), aorta (Ao), blood pressure (diastolic), left ventricular (LV) size, inter ventricular septum (IVS), ejection fraction (EF), tricuspid valve, left ventricular posterior wall (LVPW), left atrium (LA), and pulmonary velocity as input. The yield field distinguishes the chances_of coronary illness in the patient.

12.1 INTRODUCTION

Cardiovascular illness is a sort of genuine well-being endangering and regular happening ailment. The World Health Organization (WHO) has evaluated that 12 million deaths happen around the world consistently because of the cardiovascular illness. Advances in the field of medication in the course of recent decades empowered the ID of hazard factors that may contribute toward the cardiovascular maladies. The most notable reason behind coronary sickness is narrowing or blockage of the coronary veins, the veins that continuously send blood to the heart itself. This is called coronary hallway disease and happens bit by bit after some time. It's the major reason people have heart attacks.[1]

A blockage that is treated within a couple of hours causes the affected heart muscle to bite the dust. During around 30% of all heart attacks, the patient encounters no manifestations.[3] Nonetheless, unquestionable indications of the attack stay in the circulation system for quite a long time. Medicinal analysis is a significant yet convoluted undertaking that ought to be performed precisely and proficiently and its mechanization would be extremely helpful. All specialists are tragically not similarly gifted in each subforte and they are in numerous spots a rare asset.[4] A framework for mechanized medicinal analysis would upgrade therapeutic consideration and diminish costs.

With such a significant number of components to investigate for a finding of heart attacks, doctors for the most part make a conclusion by assessing patients current test outcomes. Past conclusions made on different patients with similar outcomes are likewise analyzed by doctors. These intricate methods are difficult. In this manner, a doctor must be experienced and profoundly talented to analyze heart attacks in a patient.[6] In this way, the push to use learning and experience of different specialists and clinical screening data of patients assembled in databases to support the assurance method is seen as a profitable framework that is the[2] mix of clinical choice help with PC-based patient records which could diminish medicinal mistakes, upgrade quiet well-being, decline undesirable practice variety, and improve persistent result.[8]

As of now, clinic data frameworks utilizing choice emotionally supportive networks have various apparatuses accessible to acquire information, yet they are limited. These instruments can basically answer some essential request like perceiving the male patients who are under 20 years old, and single who have been treated for heart attack. Nevertheless, they are not prepared to answer complex request given patient records, anticipating the probability of patients getting a coronary illness. Clinical decisions are much of the time made subject to pros' impulses and heuristics experience rather than on the rich data concealed in the database.[9]

They lead to unwanted tendencies, goofs, and excessive restorative costs which impact the idea of treatment provided for patients. Impelled by the need of such a system, in this chapter, a technique is prescribed to beneficially examine the coronary sickness, which results in reducing remedial errors and superfluous practice assortment, lessening indicative time, and upgrading quiet well-being and fulfillment.[12]

A major test confronting human services associations (emergency clinics, therapeutic focuses) is the arrangement of value administrations at reasonable expenses. Quality administration infers diagnosing patients accurately and overseeing medicines that are compelling. Poor clinical choices can prompt shocking results which are subsequently unsuitable. Medical clinics should likewise limit the expense of tests of clinical. They can accomplish these outcomes by utilizing fitting PC-based data and choose emotionally supportive networks. Emergency clinics and centers collect an immense measure of patient information throughout the years. These information give a premise to the investigation of hazard factors for some ailments. For instance, we can foresee the degree of heart attack to discover examples related with coronary illness.[16]

12.2 SYSTEM DESIGN

Adaptive neuro-fuzzy inference system (ANFIS) can be adequately executed for a given data/yield task and in this way it is engaging for a few application purposes. It has been successfully associated in different zones. The essential kind of NFS we apply for our data portrayal issue is alleged ANFIS model which hybridizes an ANN.[10] That is, the ANFIS model consolidates the ANN and fuzzy inference system (FIS) instruments into a compound inferring that there are no restrictions to isolate the individual features of ANN and FIS. Network-based fuzzy inference (ANFIS) is a blend of two delicate figuring strategies for ANN and fluffy rationale. Fluffy rationale can change the subjective parts of human learning and experiences into the procedure of exact quantitative examination. Be that as it may, it doesn't have a characterized strategy that can be utilized as a guide during the time spent change and human idea into principle base FIS, and it likewise sets aside a serious in length effort to alter the membership functions (MFs). Unlike ANN, it has a higher capacity in the learning procedure to adjust to its condition. Accordingly, the ANN can be utilized to naturally modify the MFs and lessen the pace of mistakes in the assurance of guidelines in fluffy rationale.[15]

ANFIS or flexible neuro cushy inference structure is a class of adaptable frameworks that are basically equivalent to soft enlistment systems. Cross breed learning estimation is used for our ANFIS model to recognize parameters. There are two parameters that go for cross breed count, forward pass and invert pass. After presentation of reason parameters, in the forward pass, a center yields push forward until layer four and the resulting parameters are resolved with least square estimate (LSE) by then goof measure is resolved for each center point.[13] In the backward pass, the slip-up sign passes on in turn around to invigorate premise parameters with point drop. MATLAB writing computer programs are used to make ANFIS model. In the essential stage, it is imperative to make a feathery inference system for the heart illnesses. To make a hidden course of action of enlistment limits, we used grid partition methodology. At the beginning of setting up, this system isolates the data space into rectangular subspaces using center point paralleled portion subject to predefined number of enlistment limits and their sorts in every estimation. Four elements were used with their enlistment limits. We used Gaussian investment work for all of the data factors.[15] An ANFIS is, in a general sense, an ANN that is for all intents and purposes tantamount to a first demand Sugeno style FIS. Consistently, there are six layers in an ANFIS model; one data layer, four covered layers, and one yield layer. Each layer plays out a particular task to propel the sign as shown in Figure 12.1.

The diagram of prediction of heart disease diagnosis system using ANFIS is computer-based automatic medical diagnosis system as illustrated in Figure 12.2. In view of the standard of ANFIS, this examination has five procedures starting with information gathering, property and information determination, information introduction, demonstrating, and assessment.[17]

a. Data Collection

Database description and preprocessing heart strike data set is gained from ESI facility Coimbatore. The data from 50 patients are used for proposed work. This database contains 13 qualities. The qualities, for example, name, age, sex, blood pressure (systolic), left ventricular posterior wall

(LVPW), blood pressure (diastolic), left ventricular (LV) size, inter ventricular septum (IVS), left atrium (LA), ejection fraction (EF), aorta (Ao), tricuspid valve, pulmonary velocity. The digitized information has 20 ordinary and 30 unusual cases. Information preprocessing is done at first to remove fundamental information and after that these information ought to be changed over into the configuration important for the forecast of hazard level. Because of the superfluous data in the coronary illness data sets, the first crude information can't be straightforwardly utilized for expectation. Thus, in information preprocessing stage, crude information should be cleaned, broken down, and changed for further advances. Cleaning and separating of the information collection is done to expel copied 29 records, standardize qualities, represent missing information, and remove immaterial information.[16]

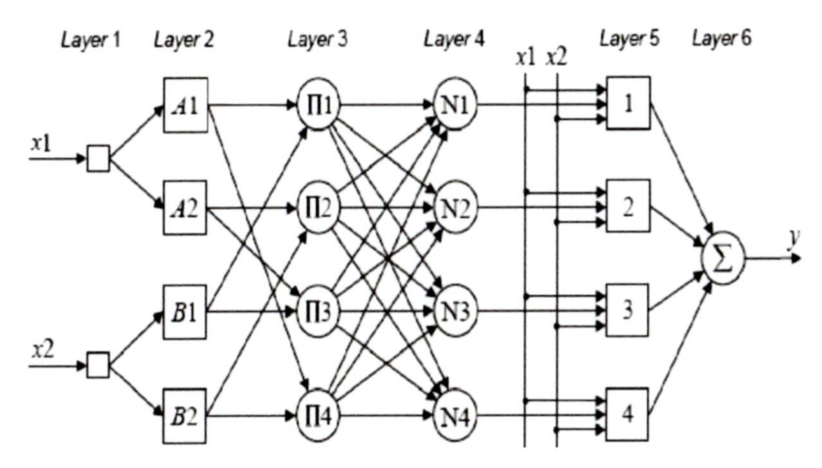

FIGURE 12.1 ANFIS architecture.

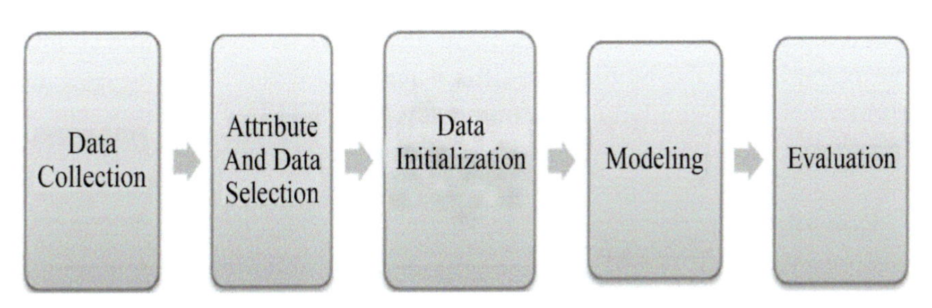

FIGURE 12.2 Workflow of ANFIS.

b. Attribute and Data Selection

Highlight determination is the way toward choosing a subset of important highlights to be utilized in building a classifier. It improves the determination execution and gives a quicker classifier. The component determination procedure cannot just diminish the expense of acknowledgment by decreasing the quantity of highlights that should be gathered, yet in addition improve the arrangement precision of the framework.[3]

c. Data Instatement

The data set is isolated into two sections that is 40% of the information is utilized for preparing and 60% is utilized for testing .Using a given information/yield information collection, the tool kit works with ANFIS to train and introduce a fluffy derivation framework (FIS) whose participation work parameters are balanced utilizing a back proliferation calculation in blend with a least squares kind of technique. This alteration enables your fluffy frameworks to gain from the information they are displaying.[6]

d. Modeling

The exhibiting approach used by ANFIS resembles various structure unmistakable confirmation strategies. An adaptable neuro fleecy finding structure or flexible framework-based soft inferring system (ANFIS) is a kind of phony neural framework that relies upon Takagi–Sugeno-cushioned determination structure. Since it arranges both neural frameworks and cushy method of reasoning norms, it can get the benefits of both in a lone structure. Its structure identifies with a great agreement of cushioned IF–THEN combination that have learning ability to vague nonlinear limits. Model endorsement is the system by which the data vectors from information/yield enlightening accumulations on which the FIS that are not ready are displayed against the ready FIS model to see how well the FIS model predicts the instructive gathering yield. We use an endorsement instructive list to check and control the potential for the model over-fitting the data. When checking data are shown to ANFIS similarly as getting ready data, the FIS model is picked to have parameters related with the base checking data model misstep.[14]

e. Evaluation

After the preparation information and creating the underlying FIS structure, we can begin testing the FIS and assess the consequence of the framework that individual is sound or not. It is possible to find a point at which level of coronary illness is the individual suffering from. Evaluation is the strategy by which the data vectors from data/yield enlightening records on which the FIS was not readied, are shown to the readied FIS model, to see how well the FIS model predicts the relating instructive file yield regards.[11]

12.3 RESULTS AND DISCUSSIONS

Heart disease prediction is simulated in MATLAB R2013a with a database that includes data set of 50 patients. The FCM utilizes fluffy parceling to such an extent that an information point can have a place with all gatherings with various enrollment reviews somewhere in the range of 0 and 1. By FCM dependent on the circulatory strain rate and pulse esteems, the figures are assembled. Figure 12.3 indicates how the bunching happens by tolerating the contributions from exceed expectations sheet.

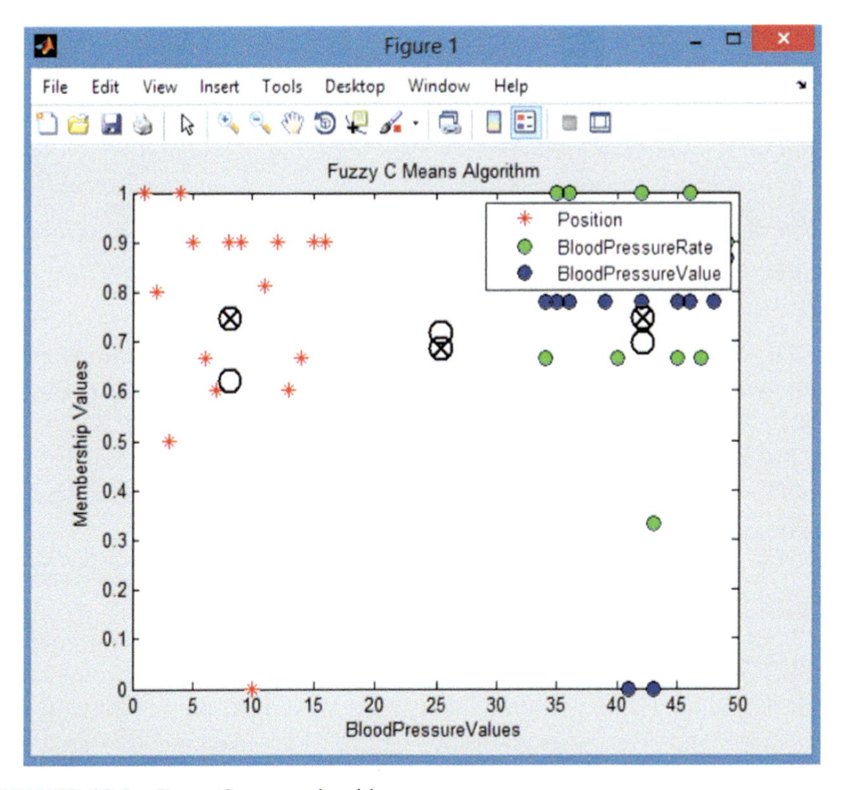

FIGURE 12.3 Fuzzy C means algorithm.

There are 15 fuzzy rules that are mainly designed based on clusters. The number of clusters taken for fuzzy rule is 4. Every one of the guidelines in the fluffy induction framework incorporates all the seven factors. These rules are shown in Figure 12.4.

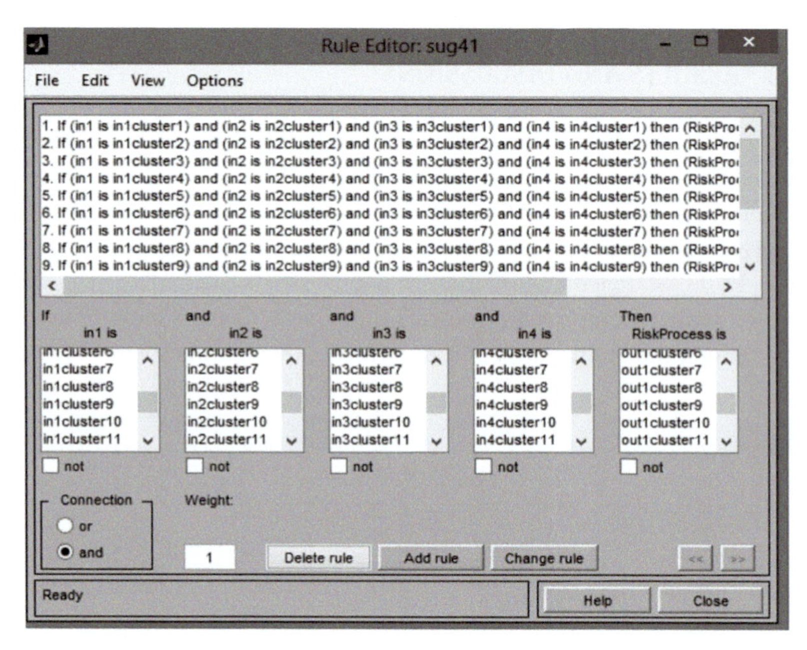

FIGURE 12.4 Fuzzy rules.

The surface watcher is a three-dimensional bend that exhibits the mapping with two information parameters to one yield for acquiring coronary illness degree. It is shown in Figure 12.5.

Figure 12.6 shows the ANFIS model structure with the membership functions in the input layer, hidden layer, and the output is produced based on the rules.

The outputs displayed on the command window is based on the heart disease prediction rate as shown in Figure 12.7 and Figure 12.8.

Figure 12.9 shows the feature extraction between Fuzzy and ANFIS It likewise demonstrates how ANFIS strategy predicts preferred outcomes over the fluffy. The expectation rate is high in ANFIS and it requires some investment and gives better precision. Table 12.1 demonstrates the outcomes for four patients from the informational collection as ordinary and irregular patient dependent on systolic pulse (mm/Hg) and Ao size (cm).

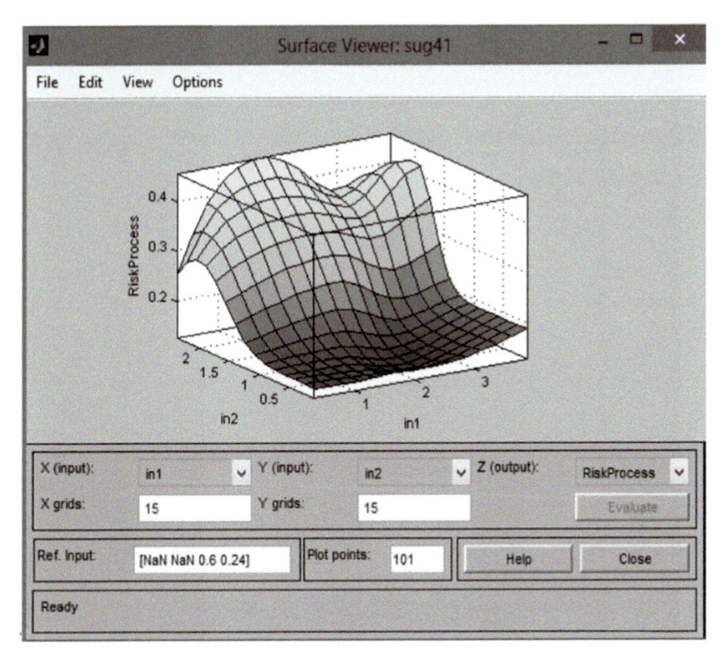

FIGURE 12.5 Surface viewer based on clusters.

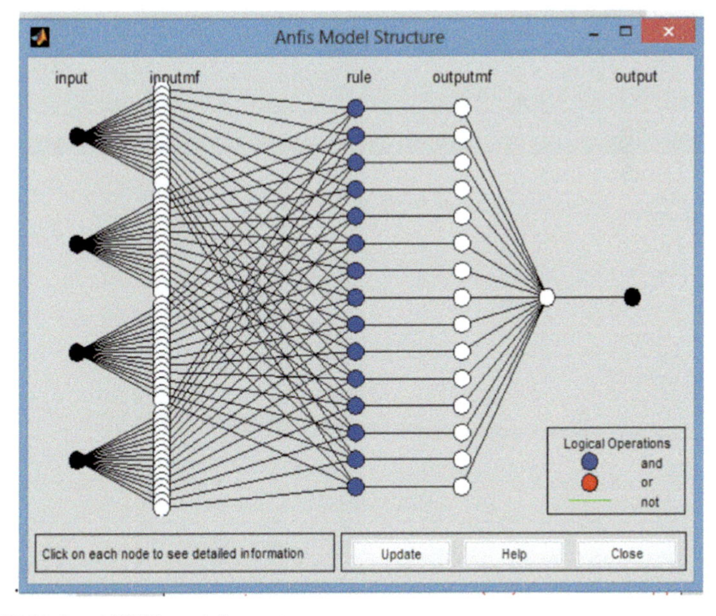

FIGURE 12.6 ANFIS model structure.

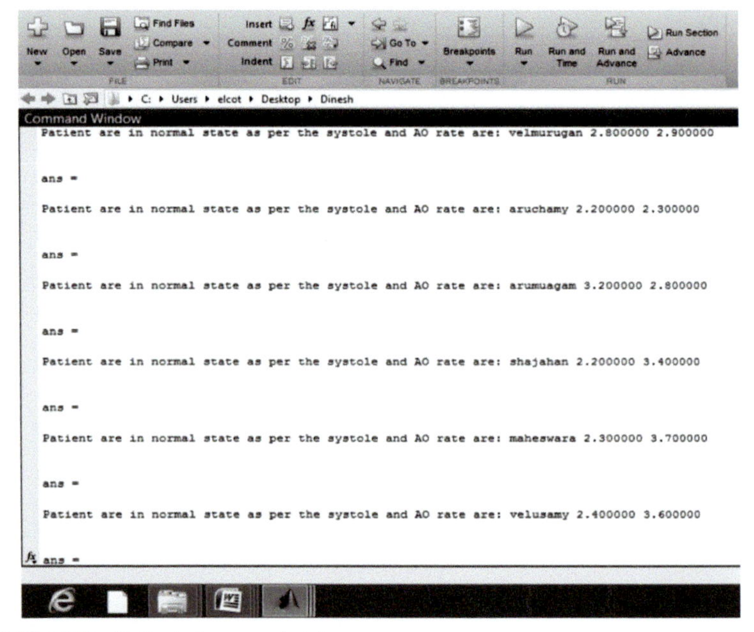

FIGURE 12.7 Patients in normal state.

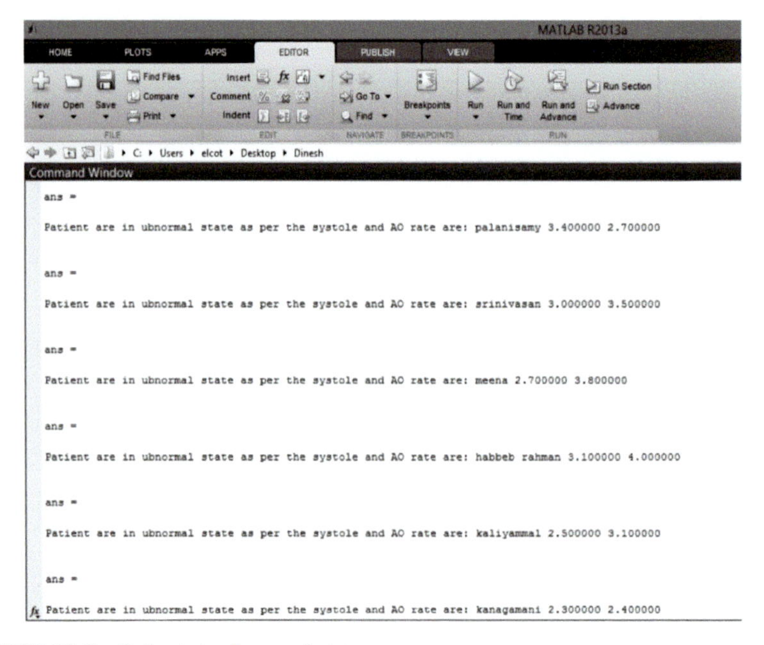

FIGURE 12.8 Patients in abnormal state.

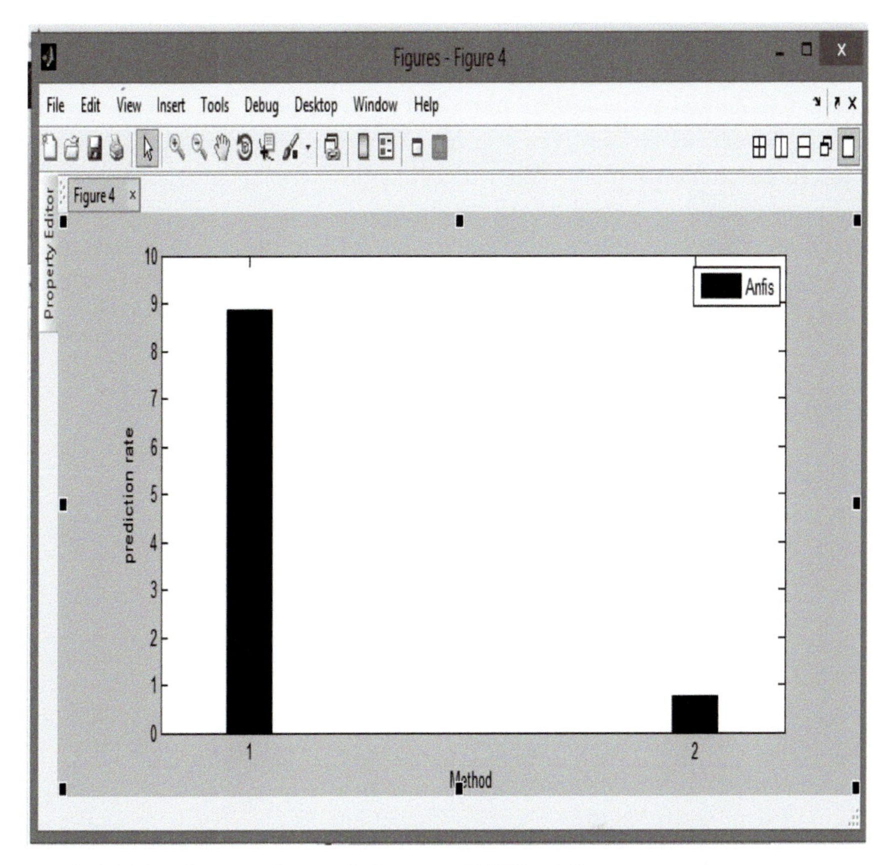

FIGURE 12.9 Comparative results between ANFIS and Fuzzy.

TABLE 12.1 Prediction Rate for Normal Versus Abnormal Patients.

Name	Systolic blood pressure	Aorta size	Result
Velmurugan	130	2.8	Normal
Aruchamy	120	2.2	Normal
Palanisamy	140	3.4	Abnormal
Sreenivasan	170	3.0	Abnormal

12.4 CONCLUSION

The extensive developing of cardiovascular infection and its belongings and intricacies, conclusion of coronary illness by utilizing AI strategies is one of the difficulties in the well-being field. In this venture, analysis of coronary illness dependent on artificial neuro fuzzy interface system is dissected. Actually, the ANFIS plays out an imperative job for expectation of sicknesses in restorative industry. The neuro-fuzzy-based inference system has been used for plan which has both the advantages of neural framework and soft method of reasoning. The data set is preprocessed and superfluous information is evacuated. The information is entered in the exceed expectations sheet and imported in MATLAB. The coronary illness is anticipated for the patients utilizing ANFIS calculation.

KEYWORDS

- **artificial intelligence**
- **LVPW**
- **LA**
- **ANFIS**
- **neural networks**
- **BPNN**

REFERENCES

1. Al-Milli, N. Back-propagation NN for Prediction of Heart Disease. *J. Theor. Appl. Inforton Tech.* **2013,** *56*(1), 1–6.
2. Ansari, A. Q. Auto Diagnosis of Heart Problem using Neuro-Fuzzy Integrated System. *Inf. Commun. Technol.* **2011,** 1383–1388.
3. Udayakumar, E.; Santhi, S.; Vetrivelan, P. An Investigation of Bayes Algorithm and Neural Networks for Identifying the Breast Cancer. *Indian J. Med. Pediatr. Oncol.* **2017,** *38*(3), 340–344.
4. Santhi, S.; Udayakumar, E.; Vetrivelan, P. An Identification of Efficient Vessel Feature for Endoscopic Analysis. *Res. J. Pharm. Tech.* **2017,** 10(8), 2633–2636.
5. Alqudah, A. A.; Abushariah Mohammad, A. M.; Adwan, O. Y.; Yousef Rana, M. M. Automatic Heart Disease Diagnosis System of Artificial Neural Network (ANN) and Neuro based Fuzzy Inference System. *J. Software Eng. App.* **2014,** *7*, 1055–1064.
6. Bhuvaneswari, R.; Kalaiselvi, K. Bayesian Class Method of Healthcare Applications. *Int. J. Comput. Teleco.* **2012,** *3*(1), 106–112.
7. Udayakumar, E.; Santhi, S.; Vetrivelan, P. *Principle and Application of Image Process Reconstructing of Cortical Surface in MR Images*; Scholars' Press, 2018; pp 1–101.

8. Kavitha, K. S.; Ramakrishnan, K. V.; Singh, M. K. Model and Design of Evolutionary Neural Network for Heart Problem Detection. *Int. J. Comput.* **2010,** *7*(5), 1–9.
9. Kumar, A. S. Diagnosis of Heart Problem of Advanced Fuzzy Mechanism. *Int. J. Sci. Inform. Tech.* **2013,** *2*, 22–30.
10. Gowrisankar, R.; Udayakumar, E., Santhi, S., Shivkumar, S.; Kumar G, S. An Unified Reeb Analysis for Cortical Surface Reconstruction of MRI Images. *Biomed. Pharmacol. J.* **2017,** *10*(2), 939–945.
11. Roopashree, A.; Udayakumar, E.; Srihari, K.; Rajesh, S.; Vaishnavi, R. Certain Investigation on Pathologies in Brain Images using MRI Slicing. *Middle-East J. Sci. Res.* **2015,** *23*, 1076–1084.
12. Kumari, M.; Godara, S. Study of Data Mining Classify in Cardio Disease Prediction. *Int. J. Comput. Tech.* **2011,** *2*(2), 304–308.
13. Loganathan, C.; Girija, K. V. Hybrid Learning for Neuro Fuzzy Based Inference System. *Int. J. Eng. Sci.* **2013,** *2*(11), 06–13.
14. Milli, N. A. Back Propagation Neural Network for Prediction of Heart Problem. *J. Theor. Appl. Inf.* **2013,** *56*, 10.
15. Udayakumar, E.; Vetrivelan, P. *Retinal Image Analysis using Neural Network and Clustering Algorithms*; Scholars' Press, 2018; pp 1–77.
16. Sayad, A. T.; Halkarnikar, P. P. Diagnosis of Heart Disease using Neural Network Approach. *Int. J. Adv. Sci. Eng. Tech.* **2014,** *2*(3).
17. Santhi, S.; Udayakumar, E.; Gowrishankar, R.; Ramesh, C.; Gowthaman, T. Region Based Image Segmenting for Newborn Brain MRI. *Biotechnol. Indian J.* **2016,** *12*(12), 1–8.

Index

B

C

D

E

F

G